Software Takes Command

INTERNATIONAL TEXTS IN CRITICAL MEDIA AESTHETICS

VOLUME #5

Software Takes Command

LEV MANOVICH

BLOOMSBURY
NEW YORK • LONDON • NEW DELHI • SYDNEY

Bloomsbury Academic
An imprint of Bloomsbury Publishing Inc

1385 Broadway
New York
NY 10018
USA

50 Bedford Square
London
WC1B 3DP
UK

www.bloomsbury.com

Bloomsbury is a registered trade mark of Bloomsbury Publishing Plc

First published 2013
Reprinted 2013

Library of Congress Cataloging-in-Publication Data
Manovich, Lev.
Software takes command: extending the language of new media/by Lev Manovich.
pages cm – (International texts in critical media aesthetics)
Includes bibliographical references and index.
ISBN 978-1-6235-6745-3 (pbk.: alk. paper) -- ISBN 978-1-6235-6817-7 (hardcover: alk. paper) 1. Computer software--Social aspects. 2. Social media. 3. Computers and civilization. 4. Mass media--Technological innovations. 5. Computer graphics. I. Title.
QA76.9.C66M3625 2013
006.7–dc23

2013002685

ISBN: HB: 978-1-6235-6817-7
PB: 978-1-6235-6745-3
ePDF: 978-1-6235-6672-2
ePUB: 978-1-6235-6261-8

Typeset by Fakenham Prepress Solutions, Fakenham, Norfolk NR21 8NN
Printed and bound in the United States of America

To Hyunjoo

CONTENTS

Acknowledgments ix
Introduction 1
Understanding media 1
Software, or the engine of contemporary societies 6
What is software studies? 10
Cultural software 20
Media applications 24
From documents to performances 33
Why the history of cultural software does not exist 39
Summary of the book's narrative 43

PART 1 Inventing media software 53

1 Alan Kay's universal media machine 55
Appearance versus function 55
"Simulation is the central notion of the Dynabook" 64
The permanent extendibility 91
The computer as a metamedium 101

2 Understanding metamedia 107
The building blocks 107
Media-independent *vs.* media-specific techniques 113
Inside Photoshop 124
There is only software 147

PART 2 Hybridization and evolution 159

3 Hybridization 161

Hybridity *vs*. multimedia 161
The evolution of a computer metamedium 176
Hybridity: examples 184
Strategies of hybridization 195

4 Soft evolution 199

Algorithms and data structures 199
What is a "medium"? 204
The metamedium or the monomedium? 225
The evolution of media species 233

PART 3 Software in action 241

5 Media design 243

After Effects and the invisible revolution 243
The aesthetics of hybridity 254
Deep remixability 267
Layers, transparency, compositing 277
After Effects interface: from "time-based" to "composition-based" 282
3D space as a media design platform 289
Import/export: design workflow 296
Variable form 307
Amplification 323

Conclusion 329

Software, hardware, and social media 329
Media after software 335
Software epistemology 337

Index 343

ACKNOWLEDGMENTS

The ideas and arguments in this book are the result of the author's interactions with hundreds of people over many years: students in classes, presenters at conferences, colleagues over email. I especially want to thank everybody who responded to my posts related to the ideas in this book on Twitter and Facebook as I was working on it from 2007 to 2012. They asked provocative questions, told me about relevant resources every time I asked, and encouraged me to go forward by asking when the book will be published.

The following people at a number of institutions played particularly key roles in the book's evolution and publication, and I would like to thank them individually (they are listed alphabetically by institution):

Bloomsbury Academic (the book publisher):
Katie Gallof, Acquisitions Editor, Film and Media Studies.
Jennifer Laing, Copy-editing.
Francisco J. Ricardo, Editor, International Texts in Critical Media Aesthetics.
Clare Turner, Lead Designer.

Software Studies Initiative (my lab established in 2007 at Calit2):
Staff:
Jeremy Douglass, Post-doctoral researcher, 2008–2012 (now Assistant Professor, English Department, UCSB).
Jay Chow, stuff member, 2012– (design, visualization, and programming).

Collaborators:
Benjamin Bratton, Associate Professor, Visual Arts, UCSD.
Elisabeth Losh, Director of Academic Programs, Sixth College, UCSD.

California Institute for Telecommunication and Information (Calit2):
Hector Bracho, media services.
Doug Ramsey, Director of Communications.
Ramesh Rao, Director, UCSD Division, Calit2.
Larry Smarr, Director, Calit2.

Center for Research in Computing and the Arts (CRCA):
Sheldon Brown, Director; Professor, Visual Arts, UCSD.
Lourdes Guardiano-Durkin, MSO.
Todd Margolis, Technical Director.

The Graduate Center, City University of New York (CUNY):
Matthew Gold, Associate Professor; Director, CUNY Academic Commons.
Tanya Domi, Director of Media Relations.
Chase Robinson, Provost and Senior Vice-Chancellor.
Jane Trombley, Executive Director for Communication and Marketing.

Software Studies Series (The MIT Press):
Matthew Fuller and Noah Wardrip-Fruin (series co-editors).
Douglass Sery, Editor, New Media, Game Studies, Design.

Finally, I want to add special thanks to Larry Smarr, Director of California Institute for Telecommunication and Information who invited me to participate in the Institute activities, helped to start my lab, and made it possible for me to work with the next generation of computing technologies being invented at Calit2 and the people inventing them.

Large parts of the book were written and edited in my favorite cafes and hotel lobbies, and I would like to thank their staff:

The Standard, West Hollywood, California.
Mondrian, West Hollywood, California.
L'Auberge Del Mar, California.
Del Mar Plaza, Del Mar, California.
Starbucks, Del Mar, California.
Sheraton Tribeca Hotel, New York.

Micki Kaufman did a great job of proofreading the manuscript at the last moment and catching many mistakes.

The book cover uses a part of a visualization created by William Huber using ImageJ software and our custom plug-ins. The visualization consists of 22,500 frames sampled at 1 frames per 3 seconds from a 62.5 hour video of the complete game play.

The book was written on Apple laptops (MacBook Pro, MacBook Air) using Microsoft Word. I used iPhone for email and social networks, and for occasional note taking.

For communication with the colleagues and the publisher, I relied on Gmail, Google Docs, Dropbox, Twitter, and Facebook.

The book illustrations were prepared by me and Jay Chow using the same popular software applications analyzed in this book: Photoshop, Illustrator, After Effects.

I am grateful to thousands of programmers and engineers who developed the software products mentioned above, and continue updating them with new features.

Introduction

Understanding media

I called my earlier book-length account of the new cultural forms enabled by computerization *The Language of New Media* (completed in 1999, it came out in 2001). By that time, the process of adoption of software-based tools in all areas of professional media production was almost complete, and "new media art" was in its heroic and vibrant stage—offering many possibilities not yet touched by commercial software and consumer electronics.

Ten years later, most media became "new media." The developments of the 1990s have been disseminated to the hundreds of millions of people who are writing blogs, uploading photos and videos to media sharing sites, and use free media authoring and editing software tools that ten years earlier would have cost tens of thousands of dollars.

Thanks to the practices pioneered by Google, the world is now used to running on web applications and services that have never been officially completed but remain forever in Beta stage. Since these applications and services run on the remote servers, they can be updated anytime without consumers having to do anything—and in fact, Google is updating its search algorithm code a few times a day. Similarly, Facebook is also updating its code daily, and sometimes it breaks. (Facebook's motto expressed in posters around its offices is "Move Fast and Break Things.") Welcome to the world of permanent change—the world that is now defined not by heavy industrial machines that change infrequently, but by software that is always in flux.

Why should humanists, social scientists, media scholars, and cultural critics care about software? Because outside of certain cultural areas such as crafts and fine art, software has replaced a diverse array of physical, mechanical, and electronic technologies used before the twenty-first century to create, store, distribute and access cultural artifacts. When you write a letter in Word (or its open source alternative), you are using software. When you are composing a blog post in Blogger or WordPress, you are using software. When you tweet, post messages on Facebook, search through billions of videos on YouTube, or read texts on Scribd, you are using software (specifically, its category referred to as "web applications" or "webware"—software which is accessed via web browsers and which resides on the servers).

And when you play a video game, explore an interactive installation in a museum, design a building, create special effects for a feature film, design a website, use a mobile phone to read a movie review or to view the actual movie, and carry out thousands of other cultural activities, in practical terms, you are doing the same thing—using software. Software has become our interface to the world, to others, to our memory and our imagination—a universal language through which the world speaks, and a universal engine on which the world runs. What electricity and the combustion engine were to the early twentieth century, software is to the early twenty-first century.

This book is concerned with "media software"—programs such as Word, PowerPoint, Photoshop, Illustrator, After Effects, Final Cut, Firefox, Blogger, WordPress, Google Earth, Maya, and 3ds Max. These programs enable creation, publishing, sharing, and remixing of images, moving image sequences, 3D designs, texts, maps, and interactive elements, as well as various combinations of these elements such as websites, interactive applications, motion graphics, virtual globes, and so on. Media software also includes web browsers such as Firefox and Chrome, email and chat programs, news readers, and other types of software applications whose primary focus is accessing media content (although they sometimes also include some authoring and editing features.)

These software tools for creating, interacting with, and sharing media represent a particular subset of application software

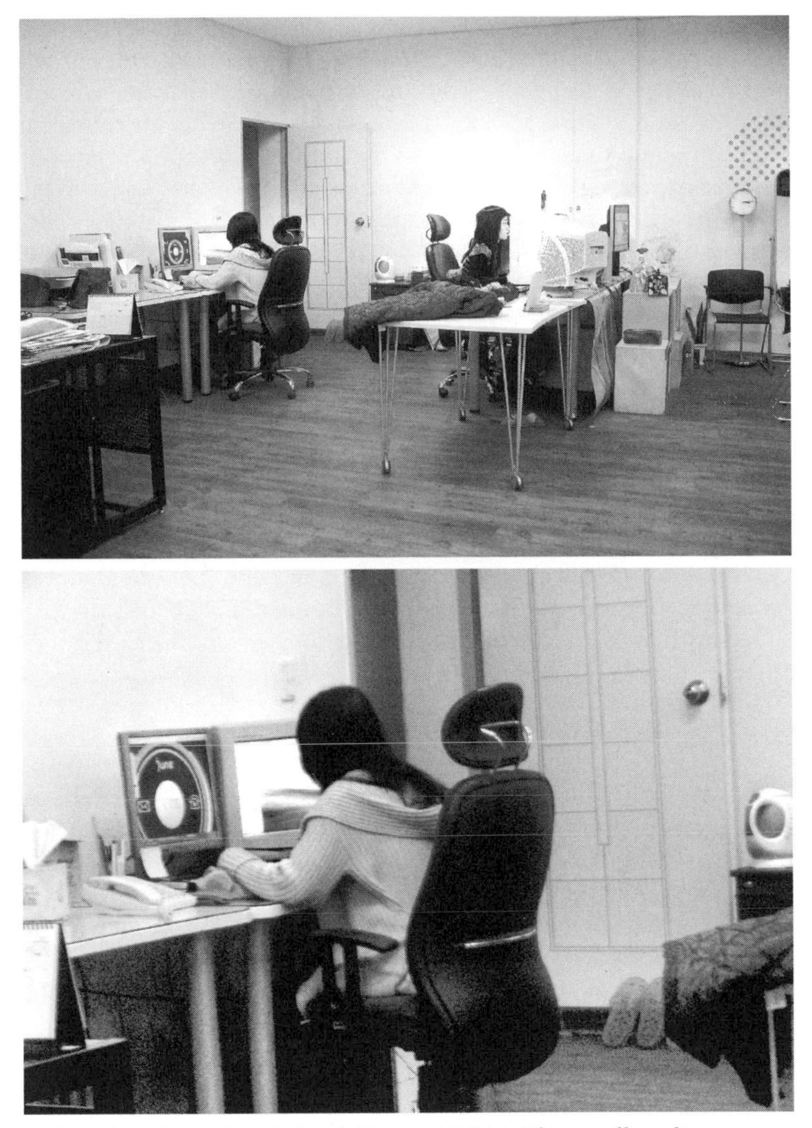

A digital studio in Seoul, South Korea, 1/2006. This small studio was responsible for the photography of all Samsung phones, to appear in its ads worldwide. In the photos we see studio staff adjusting the phone photos in Photoshop. Later these high-resolution retouched images were inserted in the Samsung TV ad, thus assuring that the product details are clearly visible.

(including web applications) in general. Given this, we may expect that all these tools inherit certain "traits" common to all contemporary software. Does this mean that regardless of whether you are working on designing a space, creating special effects for a feature film, designing a website, or making information graphics, your design process may follow a similar logic? Are there some structural features which motion graphics, graphic designs, websites, product designs, buildings, and video games share since they are all designed with software? More generally, how are interfaces and the tools of media authoring software shaping the contemporary aesthetics and visual languages of different media forms?

Behind these questions investigated in this book lies another theoretical question. This question drives the book narrative and motivates my choice of topics. What happens to the idea of a "medium" after previously media-specific tools have been simulated and extended in software? Is it still meaningful to talk about different mediums at all? Or do we now find ourselves in a new brave world of one single monomedium, or a metamedium (to borrow the term of the book's key protagonist Alan Kay)?

In short: *What is "media" after software?*

Does "media" still exist?

This book is a theoretical account of media software and its effects on the practice and the very concept of media. Over the last two decades, software has replaced most other media technologies that emerged in the nineteenth and twentieth centuries. Today it is ubiquitous and taken for granted—and yet, surprisingly, few people know about its history and the theoretical ideas behind its development. You are likely to know the names of Renaissance artists who popularized the use of linear perspective in western art (Brunelleschi, Alberti) or early twentieth-century inventors of modern film language (D. W. Griffith, Eisenstein, etc.)—but I bet you do not know where Photoshop comes from, or Word, or any other media tool you are using every day. More importantly, you probably do not know why these tools were invented in the first place.

What is the intellectual history of media software? What was the thinking and motivation of the key people and research groups they were directing—J. C. R. Licklider, Ivan Sutherland, Ted

Nelson, Douglas Engelbart, Alan Kay, Nicholas Negroponte—who between 1960 and the late 1970s created most of the concepts and practical techniques that underlie today's media applications? As I discovered—and I hope you will share my original surprise, in reading my analysis of the original texts by these people—they were as much media theoreticians as computer engineers. I will discuss their media theories and test them in view of the digital media developments in the subsequent decades. As we will see, the theoretical ideas of these people and their collaborators work very well today, helping us to better understand the contemporary software we use to create, read, view, remix, and share.

Welcome, then, to the "secret history" of our software culture—secret not because it was deliberately hidden but because until recently, excited by all the rapid transformations cultural computerization was bringing about, we did not bother to examine its origins. This book will try to convince you that such an examination is very much worth your time.

Its title pays homage to a seminal twentieth-century book *Mechanization Takes Command: a Contribution to Anonymous History* (1947) by architectural historian and critic Sigfried Giedion. In this work Giedion traces the development of mechanization in industrial society across a number of domains, including systems of hygiene and waste management, fashion, agricultural production, and food system, with separate sections of the book devoted to bread, meat, and refrigeration. Much more modest in scope, my book presents episodes from the history of "softwarization" (my neologism) of culture between 1960 and 2010, with a particular attention to media software—from the original ideas which led to its development to its current ubiquity.

My investigation is situated within a broader intellectual paradigm of "software studies." From this perspective, this book's contribution is the analysis of the ideas that eventually led to media software, and the effects of the adoption of this type of software on contemporary media design and visual culture.

Note that the category *media software* is a subset of the category *application software*; this category in its turn is a subset of the category *software*[1]—which I understand to include

[1] http://en.wikipedia.org/wiki/List_of_software_categories (July 7, 2011).

not only application software, system software, and computer programming tools, but also social network services and social media technologies.[2]

If we understand software in this extended sense, we can ask, What does it mean to live in "software society"? And what does it mean to be part of "software culture"? These are the questions the next section will take up.

Software, or the engine of contemporary societies

In the beginning of the 1990s, the most famous global brands were the companies that were in the business of producing materials or goods, or processing physical matter. Today, however, the lists of best-recognized global brands are topped with the names such as Google, Facebook, and Microsoft. (In fact, in 2007 Google became number one in the world in terms of brand recognition.) And, at least in the US, the most widely read newspapers and magazines—*New York Times*, *USA Today*, *Business Week*, etc.—feature daily news and stories about Facebook, Twitter, Apple, Google, and other IT companies.

What about other media? When I was working on the first draft of this book in 2008, I checked the business section of the CNN website. Its landing page displayed market data for just ten companies and indexes.[3] Although the list was changed daily, it was always likely to include some of the same IT brands. Let us take January 21, 2008 as an example. On that day the CNN list contained the following companies and indexes: Google, Apple, S&P 500 Index, Nasdaq Composite Index, Dow Jones Industrial

[2] Andreas Kaplan and Michael Haenlein define social media as "a group of Internet-based applications that build on the ideological and technological foundations of Web 2.0, which allows the creation and exchange of user-generated content." Andreas Kaplan and Michael Haenlein, "Users of the world, unite! The challenges and opportunities of Social Media," *Business Horizons* 53, no. 1 (January–February 2010), pp. 59–68, http://dx.doi.org/10.1016/j.bushor.2009.09.003

[3] http://money.cnn.com (January 21, 2008).

Average, Cisco Systems, General Electric, General Motors, Ford, Intel.[4]

This list is very telling. The companies that deal with physical goods and energy appear in the second part of the list: General Electric, General Motors, Ford. Right before and after these three, we see two IT companies that provide hardware: Intel makes computer chips, while Cisco makes network equipment. What about the two companies which are on top: Google and Apple? The first is in the business of information ("Google's mission is to organize the world's information and make it universally accessible and useful"[5]), while the second is making consumer electronics: phones, tablets, laptops, monitors, music players, etc. But actually, they are both making something else. And apparently, this something else is so crucial to the workings of US economy—and consequently, global world as well—that these companies almost daily appear in business news. And the major Internet companies that also appear daily in news such as Google, Facebook, Twitter, Amazon, eBay, and Yahoo, are in the same business.

This "something else" is *software*. Search engines, recommendation systems, mapping applications, blog tools, auction tools, instant messaging clients, and, of course, platforms which allow people to write new software—iOS, Android, Facebook, Windows, Linux—are in the center of the global economy, culture, social life, and, increasingly, politics. And this "cultural software"—cultural in a sense that it is directly used by hundreds of millions of people and that it carries "atoms" of culture—is only the visible part of a much larger software universe.

In *Software Society* (2003), an unrealized book proposal put together by me and Benjamin Bratton, we described the importance of software and its relative invisibility in humanities and social science research:

> Software controls the flight of a smart missile toward its target during war, adjusting its course throughout the flight. Software runs the warehouses and production lines of Amazon, Gap, Dell, and numerous other companies allowing them to assemble and dispatch material objects around the world, almost in no

---

[4] *Ibid.*

[5] http://www.google.com/about/company/ (September 23, 2012).

time. Software allows shops and supermarkets to automatically restock their shelves, as well as automatically determine which items should go on sale, for how much, and when and where in the store. Software, of course, is what organizes the Internet, routing email messages, delivering Web pages from a server, switching network traffic, assigning IP addresses, and rendering Web pages in a browser. The school and the hospital, the military base and the scientific laboratory, the airport and the city—all social, economic, and cultural systems of modern society—run on software. Software is the invisible glue that ties it all together. While various systems of modern society speak in different languages and have different goals, they all share the syntaxes of software: control statements "if then" and "while do," operators and data types (such as characters and floating point numbers), data structures such as lists, and interface conventions encompassing menus and dialog boxes.

If electricity and the combustion engine made industrial society possible, software similarly enables global information society. The "knowledge workers," the "symbol analysts," the "creative industries," and the "service industries"—none of these key economic players of the information society can exist without software.

Examples are data visualization software used by a scientist, spreadsheet software used by a financial analyst, Web design software used by a designer working for a transnational advertising agency, or reservation software used by an airline. Software is what also drives the process of globalization, allowing companies to distribute management nodes, production facilities, and storage and consumption outputs around the world. Regardless of which new dimension of contemporary existence a particular social theory of the last few decades has focused on—information society, knowledge society, or network society—all these new dimensions are enabled by software.

Paradoxically, while social scientists, philosophers, cultural critics, and media and new media theorists seem by now to cover all aspects of IT revolution, creating a number of new disciplines such as cyberculture studies, Internet studies, game studies, new media theory, digital culture, and digital humanities, the underlying engine which drives most of these subjects—software—has received comparatively little attention.

Even today, ten years later, when people are constantly interacting with and updating dozens of apps on their mobile phones and other computer devices, *software as a theoretical category* is still invisible to most academics, artists, and cultural professionals interested in IT and its cultural and social effects.

There are some important exceptions. One is the open source movement and related issues around copyright and IP that have been extensively discussed in many academic disciplines. We also see a steadily growing number of trade books about Google, Facebook, Amazon, eBay, Oracle, and other web giants. Some of these books offer insightful discussions of the software developed by these companies and the social, political, cognitive, and epistemological effects of this software. (For a good example, see John Battelle, *The Search: How Google and Its Rivals Rewrote the Rules of Business and Transformed Our Culture*.[6])

So while we are in a better situation today when we put together our proposal for *Software Society* in 2003, I feel that it is still meaningful to quote it (the only additions are the references to "social media" and "crowdsourcing"):

> If we limit critical discussions of digital culture to the notions of "open access," "peer production," "cyber," "digital," "Internet," "networks," "new media," or "social media," we will never be able to get to what is behind new representational and communication media and to understand what it really is and what it does. If we don't address software itself, we are in danger of always dealing only with its effects rather than the causes: the output that appears on a computer screen rather than the programs and social cultures that produce these outputs. "Information society," "knowledge society," "network society," "social media," "online collaboration," "crowdsourcing"—regardless of which new feature of contemporary existence a particular analysis has focused on, all these new features are enabled by software. It is time we focused on software itself.

A similar sentiment is expressed in Noah Wardrip-Fruin's *Expressive Processing* (2009) when he says in relation to books

[6] John Battelle, *The Search: How Google and Its Rivals Rewrote the Rules of Business and Transformed Our Culture* (Portfolio Trade, 2006).

about digital literature: "almost all of these have focused on what the machines of digital media look like from the outside: their output... regardless of perspective, writings on digital media almost all ignore something crucial: the actual processes that make digital media work, the computational machines that make digital media possible."[7] My book discusses what I take to be the key part of these "machines" today (because it is the only part which most users see and use directly): application software.

What is software studies?

This book aims to contribute to the developing intellectual paradigm of "software studies." What is software studies? Here are a few definitions. The first comes from my *The Language of New Media*, where, as far as I know, the terms "software studies" and "software theory" appeared for the first time. I wrote, "New media calls for a new stage in media theory whose beginnings can be traced back to the revolutionary works of Robert Innis and Marshall McLuhan of the 1950s. To understand the logic of new media we need to turn to computer science. It is there that we may expect to find the new terms, categories, and operations that characterize media that became programmable. From media studies, we move to something which can be called software studies; from media theory—to software theory."

Reading this statement today, I feel some adjustments are in order. It positions computer science as a kind of absolute truth, a given which can explain to us how culture works in software society. But computer science is itself part of culture. Therefore, I think that Software Studies has to investigate the role of software in contemporary culture, and the cultural and social forces that are shaping the development of software itself.

The book that first comprehensively demonstrated the necessity of the second approach was *New Media Reader* edited by Noah Wardrip-Fruin and Nick Montfort (The MIT Press, 2003). The publication of this groundbreaking anthology laid the framework

---

[7] Noah Wardrip-Fruin, *Expressive Processing* (Cambridge, MA: The MIT Press, 2009).

for the historical study of software as it relates to the history of culture. Although *Reader* did not explicitly use the term "software studies," it did propose a new model for how to think about software. By systematically juxtaposing important texts by pioneers of cultural computing and artists and writers active in the same historical periods, *New Media Reader* demonstrated that both belonged to the same larger epistemes. That is, often the same idea was simultaneously articulated independently by artists and scientists who were inventing cultural computing. For instance, the anthology opens with a story by Jorge Borges (1941) and an article by Vannevar Bush (1945) which both contain the idea of a massive branching structure as a better way to organize data and to capture human experience.

In February 2006 Matthew Fuller who had already published a pioneering book on software as culture (*Behind the Blip: essays on the culture of software*, 2003) organized the very first *Software Studies Workshop* at Piet Zwart Institute in Rotterdam. Introducing the workshop, Fuller wrote, "Software is often a blind spot in the theorization and study of computational and networked digital media. It is the very grounds and 'stuff' of media design. In a sense, all intellectual work is now 'software study', in that software provides its media and its context, but there are very few places where the specific nature, the materiality, of software is studied except as a matter of engineering."[8]

I completely agree with Fuller that, "all intellectual work is now 'software study.'" Yet it will take some time before the intellectuals will realize it. To help bring this change, in 2008, Matthew Fuller, Noah Wardrip-Fruin and I established the *Software Studies* book series at MIT Press. The already published books in the series are *Software Studies: A Lexicon* edited by Fuller (2008), *Expressive Processing: Digital Fictions, Computer Games, and Software Studies* by Wardrip-Fruin (2009), *Programmed Visions: Software and Memory* by Wendy Hui Kyong Chun (2011), *Code/Space: Software and Everyday Life* by Rob Kitchin and Martin Dodge (2011), and *Speaking Code: Coding as Aesthetic and Political Expression* by Geoff Cox and Alex Mclean (2012). In 2011, Fuller together with a number of UK researchers established

[8] http://pzwart.wdka.hro.nl/mdr/Seminars2/softstudworkshop (January 21, 2008).

Computational Culture, an open-access peer-reviewed journal that provides a platform for more publications and discussions.

In addition to this series, I am also happy to see a growing number of other titles written from the perspectives of platform studies, digital humanities, cyberculture, internet studies, and game studies. Many of these books contain important insights and discussions which help us better understand the roles of software. Rather than trying to list all of them, I will only provide a few examples of works which exemplify the first two of these perspectives (more will be in press by the time you are reading this). Platform studies: Nick Montfort and Ian Bogost's *Racing the Beam: The Atari Video Computer System* (2009), Jimmy Maher's *The Future Was Here: The Commodore Amiga* (2012). Digital Humanities: *Mechanisms: New Media and the Forensic Imagination (*Matthew G. Kirschenbaum, *2008), The Philosophy of Software: Code and Mediation in the Digital Age* (David Berry, 2011), *Reading Machines: Toward an Algorithmic Criticism* (Stephen Ramsay, 2011), *How We Think: Digital Media and Contemporary Technogenesis* (Katherine Hayles, 2012).[9] Also highly relevant is the first book in what may become a new area of "format studies": *MP3: The Meaning of a Format* (Jonathan Sterne, 2012).[10]

Another set of works which are relevant to understanding the roles and functioning of software systems comes from people who were trained in computer science but are also equally at home in cultural theory, philosophy, digital art, or other humanistic fields: Phoebe Sengers, Warren Sack, Fox Harrell, Michael Mateas, Paul Dourish, and Phil Agre.

Yet another relevant category of books comprises the historical studies of important labs and research groups central to the development of modern software, other key parts of information technology such as the internet, and professional practices of

---

[9] Nick Montfort and Ian Bogost, *Racing the Beam: The Atari Video Computer System* (The MIT Press, 2009); Jimmy Maher, *The Future Was Here: The Commodore Amiga* (The MIT Press, 2012); David Berry, *The Philosophy of Software: Code and Mediation in the Digital Age* (Palgrave Macmillan, 2011); Stephen Ramsay, *Reading Machines: Toward an Algorithmic Criticism* (University of Illinois Press, 2011), Katherine Hayles, *How We Think: Digital Media and Contemporary Technogenesis* (University of Chicago Press, 2012).

[10] Jonathan Sterne, *MP3: The Meaning of a Format* (Duke University Press, 2012).

software engineering such as user testing. The examples of these works listed chronologically are Katie Hafner and Matthew Lyon's *Where Wizards Stay Up Late: The Origins Of The Internet* (1998), Michael Hiltzik's *Dealers of Lightning: Xerox PARC and the Dawn of the Computer Age* (2000), Martin Campbell-Kelly's *From Airline Reservations to Sonic the Hedgehog: A History of the Software Industry* (2004), and Nathan Ensmenger's *The Computer Boys Take Over: Computers, Programmers, and the Politics of Technical Expertise* (2010).[11]

My all-time favorite book, however, remains *Tools for Thought* published by Howard Rheingold in 1985, right at the moment when domestication of computers and software starts, eventually leading to their current ubiquity. This book is organized around the key insight that computers and software are not just "technology" but rather the new *medium* in which we can think and imagine differently. Similar understanding was shared by all the heroes of this book who, with their collaborators, invented the computational "tools for thoughts"—J. C. R. Licklider, Ted Nelson, Douglass Engelbart, Bob Taylor, Alan Kay, Nicholas Negroponte. (Today many academics in humanities and social sciences still need to grasp this simple but fundamental idea. They continue to think of software as being strictly the domain of the Academic Computing Department in their universities—something which is only there to help them become more efficient, as opposed to the medium where human intellectual creativity now dwells.)

This short sketch of the intellectual landscape around software studies will be very incomplete if I do not mention the role of artists in pioneering the cultural discussions of software. Beginning around 2000, a number of artists and writers started to develop the practice of software art which included exhibitions, festivals, publishing books, and organizing online repositories of relevant works. The key figures in these developments were Amy Alexander,

[11] Katie Hafner and Matthew Lyon, *Where Wizards Stay Up Late: The Origins Of The Internet* (Simon & Schuster, 1998); Michael A. Hiltzik, *Dealers of Lightning: Xerox PARC and the Dawn of the Computer Age* (HarperBusiness, 2000); Martin Campbell-Kelly, *From Airline Reservations to Sonic the Hedgehog: A History of the Software Industry* (The MIT Press, 2004); Nathan L. Ensmenger, *The Computer Boys Take Over: Computers, Programmers, and the Politics of Technical Expertise* (The MIT Press, 2010).

Inke Arns, Adrian Ward, Geoff Cox, Florian Cramer, Matthew Fuller, Olga Goriunova, Alex McLean, Alessandro Ludovico, Pit Schultz, and Alexei Shulgin. In 2002 Christiane Paul organized CODeDOC—an exhibition of artistic code—at The Whitney Museum of American Art;[12] in 2003, the major festival of digital art Ars Electronica choose "Code" as its topic; and since 2001, the transmediale festival has included "artistic software" as one of its categories, and devoted a significant space to it in the festival's symposiums. Some of the software art projects pioneered the examination of code as the new cultural and social artifact; others offered critical commentary on commercial software practices. For example, Adrian Ward created an ironic Auto-Illustrator—"an experimental, semi-autonomous, generative software artwork and a fully functional vector graphic design application to sit alongside your existing professional graphic design utilities."

Recognizing that the bits of software studies exist across many books and art projects, Fuller writes in the Foreword to The MIT Press Software Studies book series:

> Software is deeply woven into contemporary life—economically, culturally, creatively, politically—in manners both obvious and nearly invisible. Yet while much is written about how software is used, and the activities that it supports and shapes, thinking about software itself has remained largely technical for much of its history. Increasingly, however, artists, scientists, engineers, hackers, designers, and scholars in the humanities and social sciences are finding that for the questions they face, and the things they need to build, an expanded understanding of software is necessary. For such understanding they can call upon a strand of texts in the history of computing and new media, they can take part in the rich implicit culture of software, and they also can take part in the development of an emerging, fundamentally transdisciplinary, computational literacy. These provide the foundation for Software Studies.[13]

Indeed, a number of earlier works by the leading media theorists of our times—Friedrich A. Kittler, Peter Weibel, Katherine Hayles,

---

[12] http://artport.whitney.org/commissions/codedoc/

[13] Matthew Fuller, Software Studies series introduction, http://mitpress.mit.edu/catalog/browse/browse.asp?btype=6&serid=179 (July 14, 2011).

Lawrence Lessig, Manual Castells, Alex Galloway, and others—can also be retroactively identified as belonging to software studies.[14] Therefore, I strongly believe that this paradigm has already existed for a number of years but it has not been explicitly named until a few years ago.

In his introduction to a 2006 Rotterdam workshop Fuller pointed out that "software can be seen as an object of study and an area of practice for art and design theory and the humanities, for cultural studies and science and technology studies and for an emerging reflexive strand of computer science." Since a new academic discipline can be defined either through a unique object of study, a new research method, or a combination of the two, how shall we think of software studies? Fuller's statement implies that "software" is a new object of study which should be put on the agenda of existing disciplines and which can be studied using already existing methods—for instance, actor-network theory, social semiotics, or media archaeology.

There are good reasons for supporting this perspective. I think of software as *a layer that permeates all areas of contemporary societies.* Therefore, if we want to understand contemporary techniques of *control, communication, representation, simulation, analysis, decision-making, memory, vision, writing, and interaction*, our analysis cannot be complete until we consider this software layer. Which means that all disciplines which deal with contemporary society and culture—architecture, design, art criticism, sociology, political science, art history, media studies, science and technology studies, and all others—need to account for the role of software and its effects in whatever subjects they investigate.

At the same time, the existing work in software studies already demonstrates that if we are to focus on software itself, we need new methodologies. That is, it helps to practice what one writes about. It is not accidental that all the intellectuals who have most systematically written about software's roles in society and culture have either programmed themselves or have been involved in cultural projects and practices which include writing and teaching software—for instance, Ian Bogost, Jay Bolter, Florian Cramer, Wendy Chun, Matthew Fuller, Alexander Galloway, Katherine

---

[14] See Michael Truscello, review of "Behind the Blip: Essays on the Culture of Software," *Cultural Critique* 63, Spring 2006, pp. 182–7.

Hayles, Matthew Kirschenbaum, Geert Lovink, Peter Lunenfeld, Adrian Mackenzie, Paul D. Miller, William J. Mitchell, Nick Montfort, Janet Murray, Katie Salen, Bruce Sterling, Noah Wardrip-Fruin, and Eric Zimmerman. In contrast, the scholars without this technical experience or involvement—for example, Manual Castells, Bruno Latour, Paul Virilio, and Siegfried Zielinski—have not included discussions of software in their otherwise theoretically precise and highly influential accounts of modern media and technology.

In the 2000s, the number of students in media art, design, architecture, and humanities who use programming or scripting in their work has grown substantially—at least in comparison with 1999 when I first mentioned "software studies" in The *Language of New Media*. Outside of culture and academic industries, many more people today are also writing software. To a significant extent, this is the result of new programming and scripting languages such as ActionScript, PHP, Perl, Python, and Processing. Another important factor is the publication of APIs by all major Web 2.0 companies in the middle of the 2000s. (API, or Application Programming Interface, is a code that allows other computer programs to access services offered by an application. For instance, people can use Google Maps API to embed full Google Maps on their own websites.) These programming and scripting languages and APIs did not necessarily make programming easier. Rather, they made it much more efficient. For instance, since a young designer can create an interesting work with only couple of dozen lines of code written in Processing versus writing a really long Java program, s/he is much more likely to take up programming. Similarly, if only a few lines in JavaScript allows you to integrate all the functionality offered by Google Maps into your site, this is a great motivation for beginning to work with JavaScript. Yet another reason for more people writing software today is the emergence of a massive mobile apps marketplace that, unlike the desktop market, is not dominated by a few large companies. According to informal reports in the beginning of 2012, one million programmers were creating apps for the iOS platform (iPad and iPhone) alone, and another one million were doing this for the Android platform.

In his 2006 article covering new technologies that allow people with very little or no programming experience to create new custom software (such as Ning), Martin LaMonica wrote about a future possibility of "a long tail for apps."[15] A few years later, this is exactly what happened. In September 2012, 700,000 apps were available on Apple App Store,[16] and over 600,000 Android apps on Google Play.[17]

In the article called "A Surge in Learning the Language of the Internet" (March 27, 2012), the *New York Times* reported that, "The market for night classes and online instruction in programming and Web construction, as well as for iPhone apps that teach, is booming." The article quoted Zach Sims, one of the founders of Codecademy (a web school which teaches programming though interactive lessons) who explained one of the reasons for this growing interest in learning programming and web design: "People have a genuine desire to understand the world we now live in. They do not just want to use the Web; they want to understand how it works."[18]

In spite of these impressive developments, the gap between people who can program and who cannot remains—as does the gap between professional programmers and people who just took one or two short programming classes. Clearly, today the consumer technologies for capturing and editing media are much easier to use than even the most friendly programming and scripting languages. But it does not necessarily have to stay this way. Think, for instance, of what it took to set up a photo studio and take photographs in the 1850s versus simply pressing a single button on a digital camera or a mobile phone in the 2000s. Clearly, we are very far from such simplicity in programming. But I do not see any logical reasons why programming cannot one day become equally easy.

For now, the number of people who can script and program keeps increasing. Although we are far from a true "long tail" for software, software development is gradually getting more democratized. It is, therefore, the right moment to start thinking

---

[15] Martin LaMonica, "The do-it-yourself Web emerges," *CNET News*, July 31, 2006, http://www.news.com/The-do-it-yourself-Web-emerges/2100-1032_3-6099965.html

[16] http://www.mobilestatistics.com/mobile-statistics (July 30, 2012).

[17] http://play.google.com/about/apps/ (July 30, 2012).

[18] Jenna Wortham, "A Surge in Learning the Language of the Internet," *New York Times*, March 27, 2012, http://www.nytimes.com/2012/03/28/technology/for-an-edge-on-the-internet-computer-code-gains-a-following.html

```
void setup()
{
  size(600,600);
  smooth();
  noLoop();
}

void draw()
{
  background(255);
  strokeWeight(10);
  translate(width/2,height-20);
  branch(0);
}

void branch(int depth){
  if (depth < 12) {
    line(0,0,0,-height/10);
    {
      translate(0,-height/10);
      rotate(random(-0.1,0.1));

      if (random(1.0) < 0.6){ // branching
        rotate(0.3);
        scale(0.7);
        pushMatrix();
        branch(depth + 1);
        popMatrix();
        rotate(-0.6);
        pushMatrix();
        branch(depth + 1);
        popMatrix();
      }
      else { // continue
        branch(depth);
      }
    }
  }
}
```

The complete code for tree_recursion, *a Processing sketch by Mitchell Whitelaw, 2011, http://www.openprocessing.org/sketch/8752*

Tree variations generated by tree_recursion *code.*

theoretically about how software is shaping our culture, and how it is shaped by culture in its turn. The time for "software studies" has arrived.

Cultural software

German media and literary theorist Friedrich Kittler wrote that students today should know at least two software languages: only "then they'll be able to say something about what 'culture' is at the moment."[19] Kittler himself programmed in an assembler language—which probably determined his distrust of Graphical User Interfaces and modern software applications that use these interfaces. In a classical modernist move, Kittler argued that we need to focus on the "essence" of the computer—which for Kittler meant its mathematical and logical foundations and its early history characterized by tools such as assembler languages.

This book is determined by my own history of engagement with computers as a programmer, computer animator and designer, media artist, and as a teacher. This involvement started in the early 1980s, which was the decade of procedural programming (Pascal), rather than assembly programming. It was also the decade that saw the introduction of PCs, the emergence and popularization of desktop publishing, and the use of hypertext by some literary scholars. In fact, I came to NYC from Moscow in 1981, which was the year IBM introduced their first PC. My first experience with computer graphics was in 1983–4 on Apple IIe. In 1984 I saw a Graphical User Interface in its first successful commercial implementation on an Apple Macintosh. The same year I got a job at Digital Effects, one of the first computer animation companies in the world, where I learned how to program 3D computer models and animations. In 1986 I was writing computer programs that automatically processed photographs to make them look like paintings. In January 1987 Adobe Systems shipped Illustrator,

[19] Friedrich Kittler, 'Technologies of Writing/Rewriting Technology,' *Auseinander* 1, no. 3 (Berlin, 1995), quoted in Michael Truscello, "The Birth of Software Studies: Lev Manovich and Digital Materialism," *Film-Philosophy* 7, no. 55 (December 2003), http://www.film-philosophy.com/vol7-2003/n55truscello.html

followed by Photoshop in 1989. The same year saw the release of *The Abyss*, directed by James Cameron. This movie used pioneering CGI to create the first complex virtual character. And, by Christmas of 1990, Tim Berners-Lee had already created all the components of the World Wide Web as it exists today: a web server, web pages, and a web browser.

In short, during one decade the computer moved from being a culturally invisible technology to being the new engine of culture. While the progress in hardware and Moore's Law played crucial roles in this development, even more crucial was the release of software with a Graphical User Interface (GUI) aimed at non-technical users, word processing, applications for drawing, painting, 3D modeling, animation, music composing and editing, information management, hypermedia and multimedia authoring (HyperCard, Director), and global networking (World Wide Web) With easy-to-use software in place, the stage was set for the next decade of the 1990s when most culture industries—graphic design, architecture, product design, space design, filmmaking, animation, media design, music, higher education, and culture management—gradually adapted software tools. Thus, although I first learned to program in 1975 when I was in high school in Moscow, my take on software studies has been shaped by watching how during the 1980s GUI-based software quickly put the computer in the center of culture.

If software is indeed the contemporary equivalent of the combustion engine and electricity in terms of its social effects, every type of software needs to be taken into account. We need to consider not only "visible" software used by consumers but also "grey" software, which runs all systems and processes in contemporary society. However, since I do not have personal experience writing logistics software, industrial automation software, and other "grey" software, I will be not be writing about such topics. My concern is with a particular subset of software which I used and taught in my professional life. I call it *cultural software*.

While the term "cultural software" was previously used metaphorically (see J. M. Balkin, *Cultural Software: A Theory of Ideology*, 2003), I am going to use it literally to refer to certain types of software that support actions we normally associate with "culture." These cultural actions enabled by software can be divided into a number of categories (of course we should keep in

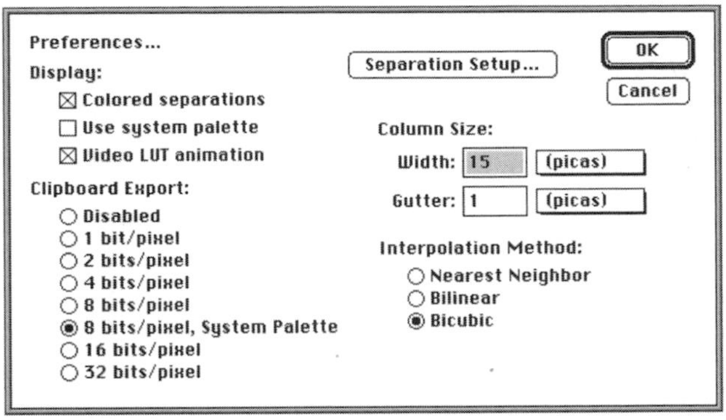

Preferences...
Display:
☒ Colored separations
☐ Use system palette
☒ Video LUT animation
Clipboard Export:
○ Disabled
○ 1 bit/pixel
○ 2 bits/pixel
○ 4 bits/pixel
○ 8 bits/pixel
◉ 8 bits/pixel, System Palette
○ 16 bits/pixel
○ 32 bits/pixel
Separation Setup...
OK
Cancel
Column Size:
Width: 15 (picas)
Gutter: 1 (picas)
Interpolation Method:
○ Nearest Neighbor
○ Bilinear
◉ Bicubic

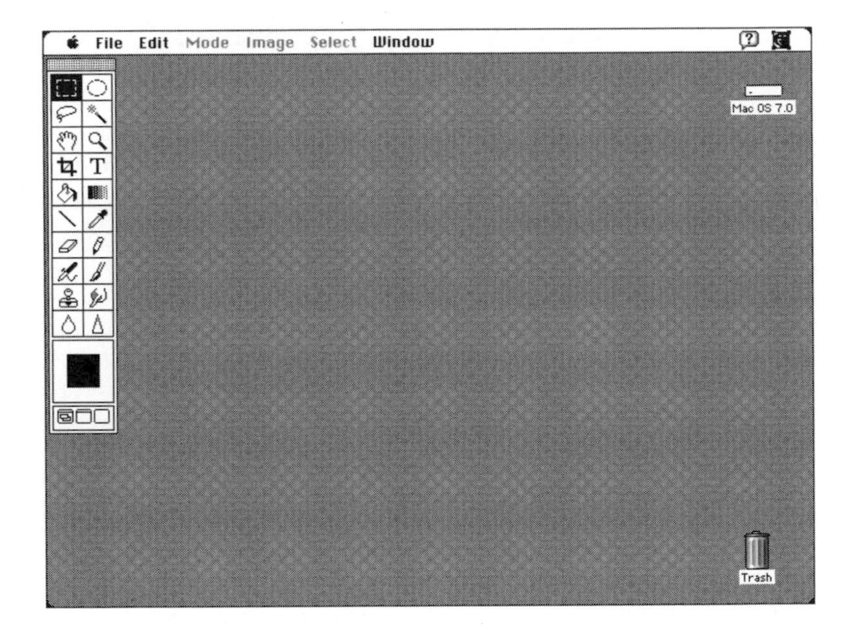

Adobe Photoshop, Macintosh version 1.0.7, 1990. Top: preferences window. Bottom: workspace.

mind that this is just one possible specific categorization system among many).

1. *Creating cultural artifacts and interactive services which contain representations, ideas, beliefs, and aesthetic values* (for instance, editing a music video, designing a package for a product, designing a website or an app).
2. *Accessing, appending, sharing, and remixing such artifacts (or their parts) online* (for instance, reading newspaper on the web, watching YouTube video, adding comments to a blog post).
3. *Creating and sharing information and knowledge online* (for instance, editing a Wikipedia article, adding places in Google Earth, including a link in a tweet).
4. *Communicating with other people* using email, instant message, voice-over IP, online text and video chat, social networking features such as wall postings, pokes, events, photo tags, notes, places, etc.
5. *Engaging in interactive cultural experiences* (for instance, playing a computer game).
6. *Participating in the online information ecology by expressing preferences and adding metadata* (for instance, automatically generating new information for Google Search whenever you use this service; clicking the "+1" button on Google+ or the "Like" button on Facebook; using the "retweet" function on Twitter).
7. *Developing software tools and services that support all these activities* (for instance, programming a library for Processing that enables sending and receiving data over the Internet;[20] writing a new plugin for Photoshop, creating a new theme for WordPress).

Technically, this software may be implemented in a variety of ways. Popular implementations (referred to in the computer industry as "software architecture") include stand-alone applications that run on the user's computing device, distributed applications (a client

---

[20] http://www.processing.org/reference/libraries/ (July 7, 2011).

running on the user's device communicates with software on the server), and peer-to-peer networks (each computer becomes both a client and a server). If all this sounds completely unfamiliar, do not worry: all you need to understand is that "cultural software" as I will use this term covers a wide range of products and network services, as opposed to only single desktop applications such as Illustrator, Photoshop or After Effects that dominated media authoring in the 1990s and 2000s. For example, social network services such as Facebook and Twitter include multiple programs and databases running on company servers (for instance, in 2007 Google was running over one million servers around the world according to one estimate[21]) and the programs (called "clients") used by people to send emails, chat, post updates, upload video, leave comments, and perform other tasks on these services. (For instance, one can access Twitter using twitter.com, or tweetdeck.com, Twitter apps for iOS, Android, and dozens of third party websites and apps.)

Media applications

Let us go through the software categories that support the first four types of cultural activities listed above in more detail.

The first category is *software for creating, editing, and organizing media content*. The examples are Microsoft Word, PowerPoint, Photoshop, Illustrator, InDesign, Final Cut, After Effects, Maya, Blender, Dreamweaver, Aperture, and other applications. This category is in the center of this book. The industry uses a number of terms to refer to this category such as "media authoring," "media editing," and "media development" but I am going to refer to this category by using a single summary term. I will simply call it *media software*.

The second category is *software for distributing, accessing, and combining (or "publishing," "sharing," and "remixing") media content on the web*. Think Firefox, Chrome, Blogger, WordPress, Tumblr, Pinterest, Gmail, Google Maps, YouTube, Vimeo and

[21] Pandia Search & Social, "Google: one million servers and counting," http://www.pandia.com/sew/481-gartner.html

other web applications and services. Obviously, the first and second categories overlap—for example, many desktop media applications allow you to upload your creations directly to popular media sharing sites, while many web applications and services include some authoring and editing functions (for example, YouTube has a built-in video editor). And blogging platforms and email clients sit right in the middle—they are used as much for publishing as for creating new content.

I will take for granted that since we all use application programs, or "apps," we have a basic understanding of this term. Similarly, I also assume that we understand what "content" refers to in digital culture, but just to be sure, here are a couple of ways to define it. We can simply list various types of media which are created, shared, and accessed with media software and the tools provided by social media and sites: texts, images, digital video, animations, 3D objects and scenes, maps, as well as various combinations of these and other media. Alternatively, we can define "content" by listing genres, for instance, web pages, tweets, Facebook updates, casual games, multiplayer online games, user-generated video, search engine results, URLs, map locations, shared bookmarks, etc.

Digital culture tends to modularize content, i.e., enabling users to create, distribute, and re-use discrete content elements—looping animations to be used as backgrounds for videos, 3D objects to be used in creating complex 3D animations, pieces of code to be used in websites and blogs, etc.[16] (This modularity parallels the fundamental principle of modern software engineering to design computer programs from small reusable parts called functions or procedures.) All such parts also qualify as "content."

Between the late 1970s and the middle of the 2000s, application programs for media editing were designed to run on a user's computer (minicomputers, PCs, scientific workstations, and later, laptops). In the next five years, companies gradually created more and more capable versions of these programs running in the "cloud." Some of these programs are available via their own websites (Google Docs, Microsoft Web Office), while others are integrated with media hosting or social media services (e.g., Photobucket image and video editor). Many applications are implemented as clients that run on mobile phones (e.g., Maps on iPhone), tablets, and TV platforms and communicate with servers and websites. Examples of such platforms are Apple's iOS,

Google's Android, and LG's Smart TV App platform. Still others are apps running on tablets such as Adobe Photoshop Touch for iPad.[22] (While at the moment of writing both web-based and mobile applications have limited editing capabilities in comparison with their desktop counterparts, this may already have changed by the time you are reading this book).

The development of mobile software platforms led to the increasing importance of certain media application types (and corresponding cultural activities) such as "media uploaders" (apps designed for uploading media content to media sharing sites). To put this differently, managing media content (for example, organizing photos in Picasa) and also "meta-managing" (i.e. managing the systems which manage it such as organizing a blogroll) have become as central to a person's cultural life as creating this content.

This book is about media software—its conceptual history, the ways it redefined the practice of media design, the aesthetics of the media being created, and creators' and users' understanding of "media." How can we place media software inside other categories and also break it into smaller categories? Let us start again with our definition, which I will rephrase here. Media software are *programs that are used to create and interact with media objects and environments*. It is a subset of the larger category of "application software"—the term which is itself in the process of changing its meaning as desktop applications (applications which run on a computer) are supplemented by mobile apps (applications running on mobile devices) and web applications (applications which consist of a web client and the software running on a server). Media software enables creation, publishing, accessing, sharing, and remixing different types of media (such as images, moving image sequences, 3D shapes, characters, and spaces, text, maps, interactive elements), as well as various projects and services which use these elements. These projects can be non-interactive (2D designs, motion graphics, film shots) or interactive (media surfaces and other interactive installations). The online services are by their very nature always interactive (websites, blogs, social networks, social media services, games, wikis, web media and app stores such as Google Play and Apple iTunes, other shopping sites,

[22] http://www.adobe.com/products/mobileapps/ (March 12, 2012).

and so on)—while a user is not always given the ability to add to or modify content, s/he always navigates and interacts with the existing content using interactive interface.

Given that today the multi-billion global culture industry is enabled by media applications, it is interesting that there is no single accepted way to classify them. The Wikipedia article on "application software" includes the categories of "media development software" and "content access software" (divided into web browsers, media players, and presentation applications).[23] This is generally useful but not completely accurate—since today most "content access software" also includes at least some media editing functions. For example, the SeaMonkey browser from Mozilla Foundation includes an HTML editor;[24] QuickTime Player can be used to cut and paste parts of video; iPhoto supports a number of photo editing operations. Conversely, in most cases "media development" (or "content creation") software such as Word or PowerPoint is used to both develop and access content. (This co-existence of authoring and access functions is an important distinguishing feature of software culture.) If we visit the websites of popular makers of these software applications such as Adobe and Autodesk, we will find that these companies may break their products by market (web, broadcast, architecture, and so on) or use sub-categories such as "consumer" and "pro." This is as good as it gets—another reason why we should focus our theoretical tools on interrogating media software.

While I will focus on media applications for creating and accessing "content" (i.e. media artifacts), cultural software also includes tools and services that are specifically designed for *communication and sharing of information and knowledge*, i.e. "social software" (categories 3–4 in my list). The examples include search engines, web browsers, blog editors, email applications, instant messaging applications, wikis, social bookmarking, social networks, virtual worlds, and prediction markets. The familiar names include Facebook, the family of Google products (Google Web search, Gmail, Google Maps, Google+, etc.), Skype, MediaWiki, and Blogger. However, since at the end of the 2000s, numerous software apps and services started to include email, post,

[23] http://en.wikipedia.org/wiki/Application_software

[24] http://www.seamonkey-project.org/

and chat functions (often via a dedicated "Share" menu), to an extent, *all software became social software.*

Of course, people do not share everything online with others—at least, not yet and not everybody. Therefore, we should also include software *tools for personal information management* such as project managers, database applications, and simple text editors or note-taking apps that are included with every computer device being sold.

These and all other categories of software shift over time. For instance, during the 2000s the boundary between "personal information" and "public information" has been reconfigured as people started to routinely place their media on media sharing sites, and also communicate with others on social networks.

In fact, the whole reason behind the existence of *social media and social networking services and hosting websites* is to erase this boundary as much as possible. By encouraging users to conduct larger parts of their social and cultural lives on their sites, these services can both sell more ads to more people and ensure the continuous growth of their user base. With more of your friends using a particular service and offering more information, media, and discussions there, you are more likely to also join that service.

As many of these services began to offer more and more advanced media editing and information management tools along with their original media hosting and communication and social networking functions, they did manage to largely erase another set of boundaries (from the PC era): those between application programs, operating system, and data. Facebook in particular was very aggressive in positioning itself as a complete "social platform" which can replace various stand-alone communication programs and services.

Until the rise of social media and the proliferation of mobile media platforms, it was possible to study media production, dissemination, and consumption as separate processes. Similarly, we could usually separate production tools, distribution technologies, and media access devices and platforms—for example, the TV studio, cameras, lighting, and editing machines (production), transmission systems (distribution), and television sets (access). Social media and cloud computing in general erase these boundaries in many cases and at the same time introduce new ones (client/server, open access/commercial). The challenge of software studies is to be able to use terms such as "content" and "software application" while always keeping in mind that the current social media/cloud computing paradigms are systematically reconfiguring the meaning of these terms.

Since creation of interactive media often involves writing some original computer code, the *programming environments* also can be considered under cultural software. Moreover, *the media interfaces* themselves—icons, folders, sounds, animations, vibrating surfaces, and touch screens—are also cultural software, since these interfaces mediate people's interactions with media and other people. I will stop here but this list can easily be extended to include additional categories of software as well.

The interface category is particularly important for this book. I am interested in how *software appears to users*—i.e. what *functions* it offers to create, share, reuse, mix, create, manage, share and communicate content, *the interfaces* used to present these functions, and *assumptions and models about a user, her/his needs, and society* encoded in these functions and their interface design.

These functions offered by an application are embedded in application commands and tools. They define what you can do with a given app, and how you can do it. This is clear; but I need to make one important point about interfaces to avoid any confusion. Many people still think that contemporary computer devices use a Graphical User Interface (GUI). In reality, the original GUI of the early 1980s (icons, folders, menus) has been gradually extended to include other media and senses (sounds, animations, and vibration feedback which may accompany user interactions on a mobile device, voice input, multi-touch gesture interfaces, etc.) This is why the term "media interface" (used in the industry) is a more accurate description of how interfaces work today. The term accurately describes interfaces of computer operating systems such as Windows and Mac OS, and mobile OS such Android and iOS; it is even applicable to interfaces of game consoles and mobile phones, as well as interactive stores[25] or museum installations which use all types of media besides graphics to communicate with the users.[26]

---

[25] For examples, see Nanika's projects for Nokia and Diesel, http://www.nanikawa.com/; Audi City in London, opened 2012.

[26] For example, see interactive installations at the Nobel Peace Center in Oslo: Nobel Chamber, Nobel Field, and Nobel Electronic Wall Papers, http://www.nobelpeacecenter.org/en/exhibitions/peace-prize-laureates/

I also need to comment on the "media/content" *vs.* "data/information/knowledge" categories used to organize my list of types of cultural software above. As with many other categories that I will use in this book, I think of them as marking the two parts of the same continuous dimension rather than as being discrete either/or boxes. A feature film is a good example of the first category, and an Excel spreadsheet represents the second category—but between such clear-cut examples, there are numerous other cases which are both. For example, if I make an information visualization of the data in the spreadsheet, this visualization now fits equally into both categories. It is still "data," but data represented in a new way which allows us to arrive at insights and "knowledge." It also becomes a piece of visual media which appeals to our senses in the same way as photographs and paintings do.

The reason that our society places these two sets of terms in opposition has to do with the histories of the media and information industries. Modern "media" is the result of the technologies and institutions which developed between the second half of the eighteenth and first half of the twentieth centuries: large-scale newspaper, magazine and book publishing, photography, cinema, radio, television and the record industry. "Data" comes from a number of separate professional fields with distinct histories: social statistics, economics, business management, and financial markets. It is only in the beginning of the twenty-first century that data leaves professional domains to become of interest to society at large. Data becomes "sexy" and "hip," with governments and cities creating their own data portals (for example, data.gov and data.gov.uk), visualizations of data entering exhibitions of major museums such as MOMA (*Design and Elastic Mind*, 2008), the computer "nerds" becoming heroes of Hollywood films (*Social Network*, 2010), and Google Analytics, Facebook, YouTube and Flickr all offering detailed data about your website or media sharing account. Of course, since media software operations (as well as any other computer processing of media for research, commercial or artistic purposes) are only possible because the computer represents *media* as *data* (discrete elements such as pixels, or equations defining vector graphics in vector files such as EPS), the development of media software and its adoption as the key media technology (discussed in this book) is an important contributor to the gradual coming together of media and data.

Software includes many other technologies and types, and computers and computer devices also perform lots of other functions besides creating and playing media. And of course, software needs hardware to run; and networks are also an essential part of our digital culture. Therefore, my focus on software applications for creating, editing, and playing media is likely to annoy some people. Not everybody uses Photoshop, Flash, Maya, and other applications to create media. A significant number of people work with media by writing their own computer programs and scripts, or modifying programs written by others. These are programmers responsible for the coding of websites, web applications, and other interactive applications, software artists, computer scientists working on the development of new algorithms, students using Processing and other high-level media programming languages, and other groups. All of them may ask me why I single out software in the form of consumer products (i.e. applications)—as opposed to the activity of programming? And what about the gradual democratization of software development and the gradual increase in the number of culture professionals and students who can program or write scripts? Should I not put my energy into promoting programming rather than explaining applications?

The reason for my choice is my commitment to understand the mainstream cultural practices rather than to emphasize (as many cultural critics do) the exceptions, no matter how progressive they may be. Although we do not have an exact number, I assume that the number of people who work in media and who can also program is tiny in comparison to the army of application users. Today, a typical professional graphic designer, film editor, product designer, architect, music artist—and certainly a typical person uploading videos to YouTube or adding photos and video on her/his blog—can neither write nor read software code. (Being able to read and modify HTML markup, or copy already pre-packaged lines of Javascript code is very different from programming.) Therefore, if we want to understand how software has already re-shaped media both conceptually and practically, we have to take a close look at the everyday tools used by the great majority of both professional and non-professional users—i.e. application software, web-based software, and, of course, mobile apps. (This book highlights the first category at the expense of the second and the third—because at this point, creation of professional media still requires applications running on a laptop or desktop, often with a significant amount of RAM and large hard drives; and also because currently web-based and mobile software are still evolving quite rapidly in contrast to desktop applications such as Photoshop and Final Cut which change only incrementally from release to release).

Any definition is likely to delight some people and to annoy others. Therefore, I also would like to address another likely objection to the way I defined the term "cultural software" (with "media software" being its subset). The term "culture" is not reducible to separate media and design "objects" which may exist as files on a computer and/or as executable software programs or scripts. It includes symbols, meanings, values, language, habits, beliefs, ideologies, rituals, religion, dress and behavioral codes, and many other material and immaterial elements and dimensions. Consequently, cultural anthropologists, linguists, sociologists, and many humanists may be annoyed at what may appear as an uncritical reduction of all these dimensions to a set of tools for creating and playing media files.

Am I saying that today "culture" is equated with a particular subset of application software and the media objects and experiences that can be created with their help? Of course not. However, what I am saying—and what I hope this book explicates in more detail—is that at the end of the twentieth century humans have added a fundamentally new dimension to everything that counts as "culture." This dimension is software in general, and application software for creating and accessing content in particular.

I am using the metaphor of a new dimension on purpose. That is, "cultural software" is not simply a new object—no matter how large and important—which has been dropped into the space which we call "culture." Thus, it would be imprecise to think of software as simply another term which we can add to the set which includes music, visual design, built spaces, dress codes, languages, food, club cultures, corporate norms, ways of talking and using a body, and so on. And while we can certainly study "the culture of software"—programming practices, values and ideologies of programmers and software companies, the cultures of Silicon Valley and Bangalore, etc.—if we only do this, we will miss the real importance of software. Like the alphabet, mathematics, printing press, combustion engine, electricity, and integrated circuits, software re-adjusts and re-shapes everything it is applied to—or at least, it has a potential to do this. Just as adding a new dimension adds a new coordinate to every point in space, "adding" software to culture changes the identity of everything that a culture is made from. (In this respect, software is a perfect example of what McLuhan meant when he wrote, the 'message of any medium or technology is the change of scale or pace or pattern that it introduces into human affairs."[27])

To summarize: our contemporary society can be characterized as a *software society* and our culture can be justifiably called a *software culture*—because today software plays a central role in shaping both the material elements and many of the immaterial structures that together make up "culture."

From documents to performances

The use of software re-configures most basic social and cultural practices and makes us rethink the concepts and theories we developed to describe them. As one example of this, consider the modern "atom" of cultural creation, transmission, and memory: a "document," i.e. some content stored in a physical form, that is delivered to consumers via physical copies (books, films, audio record), or electronic transmission (television). In software culture, we no longer have "documents," "works," "messages" or "recordings" in twentieth-century terms. Instead of fixed documents that could be analyzed by examining their structure and content (a typical move of the twentieth-century cultural analysis and theory, from Russian Formalism to Literary Darwinism), we now interact with dynamic "software performances." I use the word "performance" because what we are experiencing is constructed by software in real time. So whether we are exploring a dynamic website, playing a video game, or using an app on a mobile phone to locate particular places or friends nearby, we are engaging not with pre-defined static documents but with the dynamic outputs of a real-time computation happening on our

---

[27] Marshall McLuhan, *Understanding Media: The Extensions of Man* (New York: McGraw Hill, 1964), quoted in *New Media Reader*, Noah Wardrip-Fruin and Nick Montfort (eds) (The MIT Press, 2003), p. 203.

device and/or the server. Computer programs can use a variety of components to create these performances: design templates, files stored on a local machine, media from the databases on the network server, the real-time input from a mouse, touch screen, joystick, our moving bodies, or some other interface. Therefore, although some static documents may be involved, the final *media experience constructed by software usually does not correspond to any single static document stored in some media*. In other words, in contrast to paintings, literary works, music scores, films, industrial designs, or buildings, a critic cannot simply consult a single "file" containing all of the work's content.

Even in such seemingly simple cases as viewing a PDF document or opening a photo in a media player, we are already dealing with "software performances"—since it is software which defines the options for navigating, editing, and sharing the document, rather than the document itself. Therefore examining the PDF file or a JPEG file the way twentieth-century critics would examine a novel, a movie, or a TV program will only tell us some things about the experience we get when we interact with this document via software—but not everything. This experience is equally shaped by the interface and the tools provided by software. This is why the examination of the tools, interfaces, assumptions, concepts, and the history of cultural software—including the theories of its inventors who in the 1960s and 1970s have defined most of these concepts—is essential if we are to make sense of contemporary media.

This shift in the nature of what constitutes a media "document" also calls into question well-established cultural theories that depend on this concept. Consider the intellectual paradigm that dominated the study of media since the 1950s—the "transmission" view of culture developed in Communication Studies. Communication scholars have taken the model of information transmission formulated by Claude Shannon in his 1948 article *A Mathematical Theory of Communication* (1948)[28] and his subsequent book published with Warren Weaver in 1949,[29] and applied

[28] C. E. Shannon, "A Mathematical Theory of Communication," Bell System Technical Journal, vol. 27, pp. 379–423, 623–56, July, October, 1948, http://cm.bell-labs.com/cm/ms/what/shannonday/shannon1948.pdf

[29] Claude E. Shannon and Warren Weaver, *The Mathematical Theory of Communication* (University of Illinois Press, 1949).

its basic model of communication to mass media. The paradigm described mass communication (and sometimes culture in general) as a communication process between the authors who create and send messages and the audiences that receive them. According to this paradigm, the messages were not always fully decoded by the audiences for technical reasons (noise in transmission) or semantic reasons (they misunderstood the intended meanings).

Classical communication theory and media industries considered such partial reception a problem; in contrast, in his influential 1980 article "Encoding/decoding"[30] the founder of British Cultural Studies, Stuart Hall, argued that the same phenomenon is positive. Hall proposed that the audiences construct their own meanings from the information they receive. Rather than being a communication failure, the new meanings are active acts of intentional reinterpretation of the sent messages. But both the classical communication studies and cultural studies implicitly took for granted that the message was something complete and definite—regardless of whether it was stored in physical media (e.g. magnetic tape) or created in real time by a sender (a live TV broadcast). Thus, the receiver of communication was assumed to read all of the advertising copy, see a whole movie, or listen to the whole song and only after that s/he would interpret it, misinterpret it, assign his/her own meanings, appropriate it, remix it, etc.

While this assumption has already been challenged by the introduction of the DVR (digital video recorder) in 1999, which led to the phenomenon of *time shifting*, it simply does not apply to interactive software-driven media. The interfaces of media access applications, such as web browsers and search engines, the hyperlinked architecture of the World Wide Web, and the interfaces of particular online media services offering massive numbers of media artifacts for playback preview and/or purchase (Amazon, Google Play, iTunes, Rhapsody, Netflix, etc.), encourage people to "browse," quickly moving both *horizontally* between media (from one search result to the next, from one song to another, etc.) and vertically, *through* the media artifacts (e.g., from the contents listing of a music album to a particular track). They also made it easy to start playing/viewing media at an arbitrary point, and to

[30] Stuart Hall, "Encoding/decoding," in *Culture, Media, Language*, ed., Centre for Contemporary Cultural Studies (London: Hutchinson, 1980).

leave it at any point. In other words, the "message" that the user "receives" is not just actively "constructed" by him/her (through a cognitive interpretation) but also actively managed (defining what information s/he is receiving and how).

It is at least as important that when a user interacts with a software application that presents media content, this content often does not have any fixed finite boundaries. For instance, a user of Google Earth is likely to experience a different "Earth" every time s/he is accessing the application. Google could have updated some of the satellite photographs or added new Street Views; new 3D buildings, new layers, and new information on already existing layers were also likely to be added. Moreover, at any time a user of the application can load more geospatial data created by other users and companies by either selecting one of the options in the Add menu (Google Earth 6.2.1 interface), or directly opening a KLM file. Google Earth is a typical example of a new type of media enabled by the web—an interactive document which does not have all of its content pre-defined. Its content changes and grows over time.

In some cases this may not affect in any significant way the larger "messages" "communicated" by the software application, web service, game, or other type of interactive media. For example, Google Earth's built-in cartographic convention of representing the Earth using the General Perspective Projection (a particular map projection method of cartography) does not change when users add new content and turn on and off map layers. The "message" of this representation is always present.[31]

However, since a user of Google Earth can also add his/her own media and information to the base representation provided by the application, creating complex and media rich projects on top of existing geoinformation, Google Earth is not just a "message." It is a *platform* for users to build on. And while we can find some continuity here with the users' creative reworking of commercial media in the twentieth century—pop art and appropriation, music remixes, slash fiction and video,[32] and so on, the differences are larger than the similarities.

---

[31] http://en.wikipedia.org/wiki/Google_earth#Technical_specifications (March 14, 2012).

[32] See, for instance, Constance Penley, "Feminism, Psychoanalysis, and the Study of Popular Culture," in *Cultural Studies*, ed. Lawrence Grossberg (Routledge, 1992).

This shift from messages to platforms was in the center of the Web's transformation around 2004–6. The result was named Web 2.0. The 1990s websites presenting particular content created by others (and thus, communicating "messages") were supplemented by social networks and social media sites where the users can share, comment on, and tag their own media. The Wikipedia article on Web 2.0 describes these differences as follows: "A Web 2.0 site allows users to interact and collaborate with each other in a social media dialogue as creators (prosumers) of user-generated content in a virtual community, in contrast to websites where users (consumers) are limited to the passive viewing of content that was created for them. Examples of Web 2.0 include social networking sites, blogs, wikis, video sharing sites, hosted services, web applications, mashups and folksonomies."[33] For example, to continue with the Google Earth example, users added many types of global awareness information, including fair trade certification, Greenpeace data, and United Nations Millennium Development Goals Monitor.[34] In another example, you can incorporate Google Maps, Wikipedia, or content provided by most other large web 2.0 sites directly in your web mashup—an even more direct way of taking the content provided by web services and using it to craft your own custom platforms.

The wide adoption of Web 2.0 services along with various web-based communication tools (online discussion forums about all popular software, collaborative editing on Wikipedia, Twitter, etc.) enables quick identifications of omissions, selections, censorship and other types of "bad behavior" by software publishers—another feature which separates content distributed by web-based companies from mass media of the twentieth century. For example, every article on Wikipedia about a Web 2.0 service includes a special section about controversies, criticism, or errors.

In many cases, people can also use alternative open source equivalents of paid and locked applications. Open source and/or free software (not all free software is open source) often allow for additional ways of creating, remixing and sharing both content and new software additions. (This does not mean

---

[33] http://en.wikipedia.org/wiki/Web_2.0 (March 14, 2012).

[34] http://en.wikipedia.org/wiki/Google_earth (March 14, 2012).

that open source software always uses different assumptions and technologies than the commercial software.) For example, one can choose to use a number of alternatives to Google Maps and Google Earth—OpenStreetMap, Geocommons, WorldMap, and others which all have open source or free software licenses.[35] (Interestingly, commercial companies also often use data from such free collaboratively created systems because they contain more information than the companies' own systems. OpenStreet Map, which by early 2011 had 340,000 contributors,[36] is used by Flickr and Foursquare.[37]) A user can also examine the code of open-source software to fully understand its assumptions and key technologies.

Continuously changing and growing content of web services and sites; variety of mechanism for navigation and interaction; the abilities to add one's own content and mashup content from various sources together; architectures for collaborative authoring and editing; mechanisms for monitoring the providers—all these mechanisms clearly separate interactive networked software-driven media from twentieth-century media documents. But even when a user is working with a single local media document that is stored in a single computer file (a rather rare situation these days), such a document mediated through software interface has a different identity from a twentieth-century media document. The user's experience is only partly defined by the file's content and its organization. The user is free to navigate the document, choosing both what information to see and the sequence in which s/he is seeing it. And while "old media" (with the exception of twentieth-century broadcasting) also provided this random access, the interfaces of software-driven media players/viewers provide many additional ways for browsing media and selecting what and how to access.

For example, Adobe Acrobat can display thumbnails of every page in a PDF document; Google Earth can quickly zoom in and out from the current view; online digital libraries, databases and repositories containing scientific articles and abstracts such as the ACM

[35] http://geocommons.com, http://www.openstreetmap.org, http://worldmap.harvard.edu

[36] http://en.wikipedia.org/wiki/Counter-mapping#OpenStreetMap (March 27, 2012).

[37] http://en.wikipedia.org/wiki/OpenStreetMap#Derivations_of_OpenStreetMap_Data (March 27, 2012).

Digital Library, IEEE Xplore, PubMed, Science Direct, SciVerse Scopus, and Web of Science show articles which contain references to the one you currently selected. Most importantly, these new tools and interfaces are *not hard-wired* to the media documents themselves (such as a random access capacity of a printed book) or media access machines (such as a radio); instead they are part of the separate *software* layer. This media architecture enables easy addition of new navigation and management tools without any change to the documents themselves. For instance, with a single click, I can add sharing buttons to my blog, thus enabling new ways of circulation for its content. When I open a text document in Mac OS X Preview media viewer, I can highlight, add comments and links, draw, and add thought bubbles. Photoshop allows me to save my edits on separate "adjustment layers," without modifying the original image. And so on.

Why the history of cultural software does not exist

> "Всякое описание мира сильно отстает от его развития."
> (Translation from Russian: "Every description of the world substantially lags behind its actual development.")
>
> Тая Катюша, VJ on MTV.ru, 2008.[38]

We live in a software culture—that is, a culture where the production, distribution, and reception of most content is mediated by software. And yet, most creative professionals do not know anything about the intellectual history of software they use daily—be it Photoshop, Illustrator, GIMP, Final Cut, After Effects, Blender, Flame, Maya, MAX, or Dreamweaver.

Where does contemporary cultural software came from? How were its metaphors and techniques arrived at? And why was it developed in the first place? Currently most prominent computer and web companies have been extensively covered in media, so their history is relatively well-known (for instance, Facebook, Google, and Apple). But this is only the tip of the iceberg. The history of media authoring and editing software remains pretty much unknown. Despite the common statements that the digital revolution is at least as important as the invention of the printing press, we are largely ignorant of how the key part of this revolution—i.e., media software—was invented. When you think about this, it is unbelievable. People in the business of culture know about Gutenberg (printing press), Brunelleschi (perspective), The Lumière Brothers, Griffith and Eisenstein (cinema), Le Corbusier (modern architecture), Isadora Duncan (modern dance), and Saul Bass (motion graphics). (If you happen not to know one of these names, I am sure that you have other cultural friends who do). And yet, even today, relatively few people have heard of J. C. R. Licklider, Ivan Sutherland, Ted Nelson, Douglas Engelbart, Alan Kay, and their collaborators who, between approximately 1960 and 1978, gradually turned the computer into the cultural machine it is today.

Remarkably, the history of cultural software as a discrete category does not yet exist. What we have are a number of largely biographical books about some of the key individual figures, and research labs such as Xerox PARC or MIT Media Lab—but no comprehensive synthesis which would trace the genealogical tree of media tools. And we also do not have any detailed studies which would relate the history of cultural software to the history of media, media theory, or history of visual culture.

Modern art institutions—museums such as the MOMA and the Tate, art book publishers such as Phaidon and Rizzoli, etc.—promote the history of modern art. Hollywood is similarly proud of its own history—the stars, the directors, the cinematographers, and the classical films. So how can we understand the neglect of the history of cultural computing by our cultural institutions and computer industry itself? Why, for instance, does Silicon Valley not have a museum for cultural software? (The Computer History museum in Mountain View, California has an extensive permanent exhibition which is focused on hardware, operating systems, and programming languages—but not on the history of software.[38])

I believe that the major reason has to do with economics. Originally misunderstood and ridiculed, modern art has eventually become a legitimate investment category—in fact, by middle of the 2000s, the paintings of a number of twentieth-century artists

---

38 http://www.mtv.ru/air/vjs/taya/main.wbp (February 21, 2008).

were selling for more money than the works of the most famous classical artists. Similarly, Hollywood continues to receive profits from old movies as it reissues them in new formats (VHS, DVD, HD, Blu-ray disks, etc). What about the IT industry? It does not derive any profits from the old software—and therefore it does nothing to promote its history. Of course, contemporary versions of Microsoft Word, Adobe Photoshop, Autodesk AutoCAD, and many other popular cultural applications were built on the first versions, which often date from the 1980s, and the companies continue to benefit from the patents they filed for new technologies used in these original versions—but, in contrast to the video games from the 1980s, these early software versions are not treated as separate products which can be re-issued today. (In principle, I can imagine the software industry creating a whole new market for old software versions or applications which at some point were quite important but no longer exist today—for instance, Aldus Pagemaker. In fact, given that consumer culture systematically exploits adults' nostalgia for the cultural experiences of their teenage and youth years, it is actually surprising that early software versions were not seen as a market opportunity. If I used MacWrite and MacPaint daily in the middle of the 1980s, or Photoshop 1.0 and 2.0 in 1990–3, I think these experiences would be as much part of my "cultural genealogy" as the movies and art I saw at that time. Although I am not necessarily advocating the creation of yet another category of commercial products, if early software was widely available in simulation, it would catalyze cultural interest in software similar to the way in which wide availability of early computer games, recreated for contemporary mobile platforms, fuels the field of video game studies.

Since most theorists so far have not considered cultural software as a subject of its own, distinct from "social media," "social networks," "new media," media art," "the internet," "interactivity," and "cyberculture," we lack not only a conceptual history of media editing software but also systematic investigations of *the roles of software in media production*. For instance, how did the adoption of the popular animation and compositing application After Effects in the 1990s reshape the language of moving images? How did the adoption of Alias, Maya and other 3D packages by architectural students and young architects in the same decade similarly influence the language of architecture? What about the co-evolution of Web design tools and the aesthetics of websites—from the bare-bones HTML in 1994 to visually rich Flash-driven sites five years later, and responsive web design in the early 2010s? You will find frequent mentions and short discussions of these and similar questions in articles and conference talks, but as far as I know, there has been no book-length study about any of these subjects. Often, books on architecture, motion graphics, graphic design and other design fields will briefly discuss the importance of software tools in facilitating new possibilities and opportunities, but these discussions are not usually further developed.

In summary, a systematic examination of the connections between the workings of contemporary media software and the new communication languages in design and media (including graphic design, web design, product design, motion graphics, animation, and cinema) has not yet been undertaken. Although this book alone cannot do it all, I hope that it will provide some general models of how such connections can be teased out—as well as provide a detailed analysis of how software use redefined certain cultural areas (e.g., motion graphics and visual design).

By focusing on the theory of software for media design, this book aims to complement the work of a few other theorists that have already examined software responsible for game platforms and design (Ian Bogost, Nick Montfort), and electronic literature (Noah Wardrip-Fruin, Matthew Kirschenbaum).

In this respect, the related fields of code studies and platform studies being developed by Mark Marino,[24] Nick Montfort, Ian Bogost and others are playing a very important role. According to Marino (and I completely agree), the three fields of software studies, code studies, and game studies complement each other: "Critical code studies is an emerging field related to software studies and platform studies, but it's more closely attuned to the code itself of a program rather than the program's interface and usability (as in software studies) or its underlying hardware (as in platform studies)."[39]

[39] http://chnm2011.thatcamp.org/05/24/session-proposal-critical-codestudies/ (July 14, 2011).

Summary of the book's narrative

Between the early 1990s and the middle of the 2000s, media software has replaced most of the other media technologies that emerged in the nineteenth and twentieth centuries. Most contemporary media is created and accessed via cultural software—and yet, surprisingly, few people know about its history. What was the thinking and motivation of people who between 1960 and the late 1970s created the concepts and practical techniques that underlie today's cultural software? How does the shift to software-based production methods in the 1990s change our concepts of "media"? How have interfaces and the tools of content development software reshaped and continued to shape the aesthetics and visual languages we see in contemporary design and media? These are the key questions that I take up in this book.

My aim is not provide a comprehensive history of cultural software in general, or media authoring software in particular. Nor do I aim to discuss all the new creative techniques media software enables across dozens of cultural fields. Instead, I will trace a *particular path through this history* that will take us from 1960 to today and which will pass through some of its most crucial points. In the following I summarize this narrative and also introduce some of the key concepts developed in each part of the book.

Part 1 looks at the 1960s and 1970s. While new media theorists have spent considerable efforts in trying to understand the relationships between digital media and older physical and electronic media, the important sources—the writing and projects by Ivan Sutherland, Douglas Engelbart, Ted Nelson, Alan Kay, and other pioneers of cultural software working in these decades—still remain largely unexamined. What were their reasons for inventing the concepts and techniques that today make it possible for computers to represent, or "remediate" other media? Why did these people and their colleagues work to systematically turn a computer into a machine for media creation and manipulation? These are the questions that I take in Part 1, which explores them by focusing on the ideas and work of the key protagonist of "cultural software movement"—Alan Kay. (It is certainly possible to construct a more exclusive or an alternative history which will pay equal attention to dozens of brilliant people who worked with these people and who, together, invented all the details which form the DNA of contemporary media software—for instance, Bob Taylor, Charles Thacker, John Warnock, and others working at Xerox PARC in the 1970s; or the people who contributed to the design of the first Macintosh.[40] However, since we do not yet even have a theoretical analysis of how the ideas of the most well-known figures of the 1960s collectively changed media, this book will start with these figures, and the analysis of their theoretical writings.)

I suggest that Kay and cultural software pioneers aimed to create a particular kind of new media—rather than merely simulating the appearances of old ones. These new media use already existing representational formats as their building blocks, while adding many previously nonexistent properties. At the same time, as envisioned by Kay, these media are expandable—that is, users themselves should be able to easily add new properties, as well as to invent new media. Accordingly, Kay calls computers the first *metamedium* whose content is "a wide range of already-existing and not-yet-invented media."

The foundations necessary for the existence of such metamedium were established between the 1960s and the late 1970s. During this period, most previously available physical and electronic media were systematically simulated in software, and a number of new media were also invented. This development takes us from the very interactive design program—Ivan Sutherland's Sketchpad (1962)—to the commercial desktop applications that made software-based media authoring and design widely available to members of different creative professions and, eventually, media consumers as well—AutoCAD (1982), Word (1984), PageMaker (1985), Alias (1985), Illustrator (1987), Director (1987), Photoshop (1989), After Effects (1993), and others. (These PC applications were paralleled by much more expensive systems for professional markets such as the TV and video industries which got Paintbox in 1981, Harry in 1985, Avid in 1989, and Flame in 1992.)

So what happens next? Did Kay's theoretical formulations as articulated in 1977 accurately predict the developments of the next thirty years, or have there been new developments that his concept of "metamedium" did not account for? Today we do

---

[40] For the stories that document the inventions by dozens of people of the multiple technologies that made up the original Macintosh, see www.folklore.com

indeed use a variety of previously existing media simulated in software as well as new previously non-existent media types. Both have been continuously extended with new properties. Do these processes of invention and amplification take place at random, or do they follow particular paths? In other words, what are the key mechanisms responsible for the extension of the computer metamedium?

Parts 2 and 3 are devoted to these questions. They look at the number of different mechanisms which drove development and expansion of the computer metamedium, with the focus on the 1990s when media software was gradually adopted in all areas of professional media production. I use three different concepts to describe these developments and the new aesthetics of visual media which developed in the second part of the 1990s after the processes of adoption reached sufficient speed. These three concepts are *media hybridization*, *evolution*, and *deep remix*. Part 2 develops the theoretical analysis of this second stage of metamedium development, illustrating it with a number of examples drawn from different genres of digital media. Part 3 focuses in detail on the use of software for visual design (motion graphics and graphics design), analyzing the relationships between the new aesthetics of moving and still images and compositions, and the operations and interfaces of software used to create them such as After Effects.

I argue that in the process of the translation from physical and electronic media technologies to software, all individual techniques and tools that were previously unique to different media "met" within the same software environment. This meeting had fundamental consequences for human cultural development and for the media evolution. It disrupted and transformed the whole landscape of media technologies, the creative professions that use them, and the very concept of media itself.

Once they were simulated in a computer, previously incompatible techniques of different media begin to be combined in endless new ways, leading to new media hybrids, or, to use a biological metaphor, new "media species." As just one example among countless others, think, for instance, of the popular Google Earth application, combining techniques of traditional mapping, the concepts from the field of Geographical Information Systems (GIS), 3D computer graphics and animation, social software, search, and other elements and functions. In my view, this *ability to combine previously separate media techniques represents a fundamentally new stage in the history of human media, human semiosis, and human communication*, enabled by its "softwarization."

I describe this new stage in media evolution using the concept of *hybridity*. In the first stage, most existing media were simulated in a computer and a number of new types of media that can only be realized in a computer were invented. In the second stage, these simulated and new mediums started exchanging properties and techniques.

To distinguish these processes from more familiar remixes, I introduce the new term *deep remixability*. Normally a remix is a combination of content from a single medium (like in music remixes), or from a few mediums (like Anime Music Video works which combine content from anime and music video). However, the software production environment allows designers to remix not only the content of different media types, but also their fundamental techniques, working methods, and ways of representation and expression.

While today hybridization and deep remix can be found at work in all areas of culture where software is used, I focus on particular area to demonstrate how it functions in detail. This area is *visual design* in general, and *motion graphics* in particular. Motion graphics is a dynamic part of contemporary culture, which, as far as I know, has not yet been theoretically analyzed in detail anywhere. Although selected precedents for contemporary motion graphics can already be found in the 1950s and 1960s in the works by Saul Bass and Pablo Ferro, its exponential growth from the middle of the 1990s is directly related to the adoption of software for moving image design—specifically, After Effects software released by Adobe in 1993. Deep remixability is central to the aesthetics of motion graphics. That is, the larger proportion of motion graphics projects done today around the world derive their aesthetic effects from combining different techniques and media traditions—animation, drawing, typography photography, 3D graphics, video, etc.—in new ways. As a part of my analysis, I look at how the typical software-based production workflow in a contemporary design studio—the ways in which a project moves from one software application to another—shapes the aesthetics of motion graphics, and visual design in general.

The next major wave of computerization of culture has to do with different types of software—social networks, social media services, and apps for mobile platforms. The wave of social networks and social media started slowly, erupted in 2005–2006 (Flickr, YouTube) and continues to move forward and expand its reach. The 1990s' media revolution impacted *professional creatives*; the 2000s' media revolution affected *the rest of us*—i.e. the hundreds of millions who use Facebook, Twitter, Firefox, Safari, Google Search and Maps, Flickr, Picasa, Vimeo, Blogger, and numerous apps and services available on mobile platforms.

Because we are still in the middle of social media diffusion, with some popular social media services going out of favor and others gaining speed (for example, think of the fate of MySpace), and the "social" functionality of software still expanding, I decided that offering the detailed theoretical analysis of this new wave would be premature. (This became clear after I started editing the part about social media which I originally had in the first book draft, and realized that some of the social media services I was analyzing in detail no longer exist… .) Instead, I am focusing on tracing the fundamental developments which made possible and shaped "digital media" before its social explosion: the ideas about the computer as a machine for media generation and editing of the 1960s–1970s, their implementation in the media applications in the 1980s–1990s, and the transformation of visual media languages which quickly followed.

To be more precise, we can frame this history between 1961 and 1999. In 1961, Ivan Sutherland at MIT designed Sketchpad, which became the first computer design system shown to the public. In 1999, After Effects 4.0 introduced Premiere import,[41] Photoshop 5.5 added vector shapes,[42] and Apple showed the first version of Final Cut Pro[43]—in short, the current paradigm of interoperable media authoring and editing tools capable of creating professional media without special hardware beyond the off-the-shelf computer was finalized. And while professional media tools continued to evolve after this period, the changes so far have been incremental. Similarly, the languages of professional visual media created with this software did not change significantly after their radical transformation in the second part of the 1990s.

---

[41] http://en.wikipedia.org/wiki/After_Effects#History (July 7, 2011).
[42] http://en.wikipedia.org/wiki/Adobe_Photoshop_release_history (July 7, 2011).
[43] http://en.wikipedia.org/wiki/Final_Cut_Pro#History (July 7, 2011).

To illustrate this continuity, the examples of particular media projects that I will analyze will be drawn from both the 1990s and the 2000s. However, when discussing interfaces and commands of media applications, I will use the recent versions in order to make the discussion as relevant as possible for the software users. Accordingly, I will also take into account the addition of social media capacities in all consumer-level media software that took place at the end of the 2000s (e.g., the "share" menu in iPhoto). And because the hybridization mechanism is not limited to professional media software and professionally created media, but also plays the key role in evolution of social web software and services, I will include prominent examples of such services, for example, Google Earth.

I also need to comment on my choices of particular media software applications as examples. I have chosen to focus on the desktop applications for media authoring most widely used today—Photoshop, Illustrator, InDesign, Dreamweaver, After Effects, Final Cut, Maya, 3ds Max, Word, PowerPoint, etc. These programs exemplify different categories of media authoring software: image editing, vector graphics, page layout, web design, motion graphics, video editing, 3D modeling and animation, word processing, and presentation. I will also be making references to popular web browsers (Firefox, Chrome, Internet Explorer), blogging tools and publishing services (WordPress, Blogger), social networks (Facebook, Twitter, Google+), media sharing services (Flickr, Pinterest, YouTube, Vimeo), email services and clients (Gmail, Microsoft Outlook), web-based office suites (Google Docs), and consumer geographic information systems (Google Earth, Bing Maps). Since I am interested in how users interact with media, another key software category for this book is media players pre-installed on new computers (Windows Media Player, iTunes, QuickTime) and document viewing applications (Adobe Reader, Mac OS Preview). As some programs and web services become less popular and new ones gain market share, the list above may look somewhat different by the time you are reading this, and many applications may also fully migrate from desktop to the web—but the categories are likely to remain the same.

Because I want to make my discussions as relevant as possible to contemporary designers and artists, the names of historically important programs which are no longer popular or do not exist will only be mentioned in passing. Examples are QuarkXPress, WordPerfect, and Macromedia Director. Luckily for us, the two programs that I will analyze in detail—Photoshop (Chapter 2) and After Effects (Chapter 5)—are as popular today as they were in the 1990s.

I will not be discussing other types of digital media authoring and editing systems which were quite important in the 1980s and 1990s. Because during this period graphics capacities of personal computers were still limited, these systems ran on graphics workstations (specialized minicomputers) from Silicon Graphics or used proprietary hardware. Here are examples listed in chronological order, with the function of the system, company name and year of the first release appearing in parentheses: Paintbox (graphics for broadcast television, Quantel, 1981), Mirage (digital real-time video effects processor, Quantel, 1982),[44] Personal Visualizer (3D modeling and animation, Wavefront, 1988), Henry and Hal (effects editor and graphics and compositing systems, Quantel, 1992) Inferno and Flame (compositing for film and video, Discreet Logic, 1992).

In the middle 1990s, Flame together with the SGI workstation cost $450,000; an Inferno system cost $700,000.[45] Inferno 5 and Flame 8 introduced in 2003 had suggested list prices of $571,500 and $266,500, respectively.[46] Because of these prices, such systems were only used in television and film studios or in big video effects companies.

Today the most demanding areas of media production which involve working with massive amounts of data—feature films, feature animations, TV commercials—still rely on these systems' expensive software. While at the end of the 2000s the companies started to offer versions of these programs for PC, Macs, and Linux, today the highest-end versions still often require special hardware, and their prices are still quite high. (For example, the 2010 edition of Autodesk Flame Premium, a suite containing Smoke, Flame and

---

[44] http://en.wikipedia.org/wiki/Quantel_Mirage (August 23, 2012).

[45] "Discreet Logic Inc. History," http://www.fundinguniverse.com/company-histories/discreet-logic-inc-history/

[46] Autodesk, "Discreet Delivers inferno 5, flame 8 and flint 8," January 23, 2003, http://investors.autodesk.com/phoenix.zhtml?c=117861&p=irol-newsArticle&ID=374002

Lustre used for video editing, effects, and color grading was offered for $125,000.)

Because I do not expect a typical reader of this book to have a working experience with these expensive systems, I will not be referring to them further in this book. However, a more comprehensive history of the moving image media of the 1980s–1990s (which I hope somebody will write in the future) will definitely need to do an archeology and genealogy of these systems and their use.

Finally, one more explanation is in order. Some readers will be annoyed that I focus on commercial applications for media authoring and editing, as opposed to their open source alternatives. For instance, I discuss Photoshop rather than Gimp, and Illustrator rather than Inkscape. I love and support open source and free access, and use it for all my work. Starting in 1994, I was making all my articles available for free download on my website manovich.net. And when in 2007 I set up a research lab (www.softwarestudies.com) to start analyzing and visualizing massive large media datasets, we decided to also follow a free software/open source strategy, making the tools we develop freely available and allowing others to modify them.[47]

The reason this book focuses on commercial media authoring and editing software rather than its open source equivalents is simple. In almost all areas of software culture, people use free applications and web services. The examples include web browsers, web email, social networks, apps available for mobile devices, and programming and scripting languages. The companies are not charging for these free applications and services because they are making money in other ways (advertising, charging for extra features and services, membership fees, selling devices). However, in the case of professional tools for media authoring and editing, commercial software dominates. It is not necessarily better, but it is simply used by many more people. (For example, entering "Photoshop" and "Gimp" into Google Trends shows that since 2004, the number of searches for the former is about eight times bigger than for the latter.) Since I am interested in describing the common user experiences, and the features of media aesthetics

[47] http://lab.softwarestudies.com/p/software-for-digital-humanities.html

common to millions of works created with the most common authoring tools that are all commercial products, these are the products I choose to analyze. And when I analyze tools for media access and collaboration, I similarly choose the most popular products—which in this case includes both free software and services provided by commercial companies (Safari, Google Earth), and free open source software (Firefox).

PART ONE

Inventing media software

CHAPTER ONE

Alan Kay's universal media machine

Medium:
8.a. A specific kind of artistic technique or means of expression as determined by the materials used or the creative methods involved: the medium of lithography.
b. The materials used in a specific artistic technique: oils as a medium.

American Heritage Dictionary, 4th edition
(Houghton Mifflin, 2000)

"The best way to predict the future is to invent it."

Alan Kay

Appearance versus function

Between its invention in the mid-1940s and the arrival of PCs in the early 1980s, the digital computer was mostly used for military, scientific, and business calculations and data processing. It was not interactive. It was not designed to be used by a single person. In short, it was hardly suited for cultural creation.

As a result of a number of developments of the 1980s and 1990s—the rise of the personal computer industry, adoption of Graphical User Interfaces (GUI), the expansion of computer networks and the World Wide Web—computers moved into the cultural mainstream. Software replaced many other tools and technologies for creative professionals. It has given hundreds of millions of people the abilities to create, manipulate, sequence and share media—but has this led to the invention of fundamentally *new* forms of culture? Today media companies are busy promoting *e-books* and interactive *television*; the consumers are happily purchasing music *albums* and *feature films* distributed in digital form, as well making *photographs* and *video* with their digital cameras and cell phones; office workers are reading PDF *documents which imitate paper*.

In short, it appears that the revolution in the means of production, distribution, and access of media has not been accompanied by a similar revolution in the syntax and semantics of media. Who shall we blame for this? Shall we put the blame on the pioneers of cultural computing—J. C. R. Licklider, Ivan Sutherland, Ted Nelson, Douglas Engelbart, Seymour Paper, Nicholas Negroponte, Alan Kay, and others? Or, as Nelson and Kay themselves are eager to point out, does the problem lie with the way the industry implemented their ideas?

Before we blame the industry for bad implementation—we can always pursue this argument later if necessary—let us look into the thinking of the inventors of cultural computing themselves. For instance, what about the person who guided the development of a prototype of a modern person computer—Alan Kay?

Between 1970 and 1981 Alan Kay was working at Xerox PARC—a research center established by Xerox in Palo Alto. Building on the already accomplished work of the pioneers of cultural computing, the Learning Research Group at Xerox PARC headed by Kay, systematically articulated the paradigm and the technologies of *vernacular media computing,* as it exists today.[1]

---

[1] Kay has expressed his ideas in a few articles and a large number of interviews and public lectures. The following have been my main primary sources: Alan Kay and Adele Goldberg, *Personal Dynamic Media, IEEE Computer* 10, no. 3 (1977); Alan Kay, "The Early History of Smalltalk," The 2nd ACM SIGPLAN Conference on History of Programming Languages (New York: ACM, 1993), pp. 69–95; Alan Kay, "A Personal Computer for Children of All Ages," *Proceedings of the ACM 1972 National Conference* (Boston, 1972); Alan Kay, *Doing with Images Makes Symbols,* videotape (University Video Communications, 1987), http://archive.org/details/AlanKeyD1987/; Alan Kay, "User Interface: A Personal View," in *The Art of*

Although selected artists, filmmakers, musicians, and architects were already using computers since the 1950s, often developing their software in collaboration with computer scientists working in research labs (Bell Labs, IBM Watson Research Center, etc.) most of this software was aimed at producing only particular kinds of images, animations or music, congruent with the ideas of their authors. In addition, each program was designed to run on a particular machine. Therefore, these software programs could not function as general-purpose tools easily usable by others.

It is well known most of the key ingredients of personal computers as they exist today came out of Xerox PARC: the Graphical User Interface with overlapping windows and icons, bitmapped display, color graphics, networking via Ethernet, mouse, laser printer, and WYSIWYG ("what you see is what you get") printing. But what is equally important is that Kay and his colleagues also developed a range of applications for media manipulation and creation that also all used a graphical interface. They included a word processor, a file system, a drawing and painting program, an animation program, a music editing program, etc. Both the general user interface and the media manipulation programs were written in the same programming language, Smalltalk. While some of the applications were programmed by members of Kay's group, the users that included seventh-grade high-school students programmed others.[2] (This was consistent with the essence of Kay's vision: to provide users with a programming environment, examples of programs, and already-written general tools so the users would be able to make their own creative tools.)

When Apple introduced the first Macintosh computer in 1984, it brought the vision developed at Xerox PARC to consumers (the new computer was priced at USD $2,495). The original Macintosh 128K included a word processing and a drawing application (MacWrite and MacPaint, respectively). Within a few years these were joined by other software for creating and editing

Human-Computer Interface Design, ed. Brenda Laurel (Reading, Mass: Addison-Wesley, 1990), pp. 191–207; David Canfield Smith *et al.*, "Designing the Star user Interface," *Byte*, issue 4 (1982).

[2] Alan Kay and Adele Goldberg, "Personal Dynamic Media," in *New Media Reader*, ed. Noah Wardrip-Fruin and Nick Montfort (The MIT Press, 2003), p. 399.

different media: Word, PageMaker and VideoWorks (1985),[3] SoundEdit (1986), Freehand and Illustrator (1987), Photoshop (1990), Premiere (1991), After Effects (1993), and so on. In the early 1990s, similar functionality became available on PCs running Microsoft Windows.[4] And while Macs and PCs were at first not fast enough to offer true competition for traditional media tools and technologies (with the exception of word processing), other computer systems specifically optimized for media processing started to compete with these technologies in the 1980s. (The examples are the NeXT Workstation, produced between 1989 and 1996; Amiga, produced between 1985 and 1994; and Paintbox, first released in 1981.)

By around 1991, the new identity of a computer as a personal media editor was firmly established. (This year Apple released QuickTime, which brought video to the desktop; the same year saw the release of James Cameron's *Terminator II*, which featured pioneering computer-generated special effects). The vision developed at Xerox PARC became a reality—or rather, one important part of this vision in which the computer was turned into a personal machine for display, authoring and editing content in different media. And while in most cases Alan Kay and his collaborators were not the first to develop particular kinds of media applications—for instance, paint programs and animation programs were already written in the second part of the 1960s[5]—by implementing all of them on a single machine and giving them consistent appearance and behavior, Xerox PARC researchers established a new paradigm of media computing.

I think that I have made my case. The evidence is overwhelming. It is Alan Kay and his collaborators at PARC that we must take to task for making digital computers imitate older media. By developing easy-to-use GUI-based software to create and edit familiar media types, Kay and others appear to have locked the computer into being a simulation machine for "old media." Or, to put this in terms of Jay Bolter and Richard Grusin's influential book *Remediation: Understanding New Media* (2000), we can say that GUI-based software turned a digital computer into a "remediation machine:"

---

[3] Videoworks was renamed Director in 1987.

[4] 1982: AutoCAD; 1989: Illustrator; 1992: Photoshop, QuarkXPress.

[5] See http://sophia.javeriana.edu.co/~ochavarr/computer_graphics_history/historia/

a machine that expertly represents a range of earlier media. (Other technologies developed at PARC, such as the bitmapped color display used as the main computer screen, laser printing, and the first Page Description Language which eventually lead to Postscript, were similarly conceived to support the computer's new role as a machine for simulation of physical media.)

Bolter and Grusin define remediation as "the representation of one medium in another."[6] According to their argument, new media always remediate the old ones and therefore we should not expect that computers would function any differently. This perspective emphasizes the continuity between computational media and earlier media. Rather than being separated by different logics, all media including computers follow the same logic of remediation. The only difference between computers and other media lies in how and what they remediate. As Bolter and Grusin put this in the first chapter of their book, "What is new about digital media lies in their particular strategies for remediating television, film, photography, and painting." In another place in the same chapter they make an equally strong statement that leaves no ambiguity about their position: "We will argue that remediation is a defining characteristic of the new digital media."

If today we consider all the digital media created by both consumers and by professionals—digital photography and video shot with inexpensive cameras and cell phones, the contents of personal blogs and online journals, illustrations created in Photoshop, feature films cut on Avid, etc.—in terms of its appearance digital media indeed often looks exactly the same way as media before computers. Thus, if we limit ourselves to looking at the media surfaces, the remediation argument accurately describes much of computational media. But rather than accepting this condition as an inevitable consequence of the universal logic of remediation, we should ask why this is the case. In other words, if contemporary computational media imitates other media, how did this become possible? There was definitely nothing in the original theoretical formulations of digital computers by Turing or Von Neumann about computers imitating other media such as books, photography, or film.

[6] Jay Bolter and Richard Grusin, *Remediation: Understanding New Media* (The MIT Press, 2000).

The conceptual and technical gap which separates the first room-sized computers used by the military to calculate the shooting tables for anti-aircraft guns and crack German communication codes, and contemporary small desktops and laptops used by ordinary people to create, edit and share media is vast. The contemporary identity of a computer as a media processor took about forty years to emerge, if we count from 1949 when MIT's Lincoln Laboratory started to work on first interactive computers to 1989 when the first commercial version of Photoshop was released. *It took generations of brilliant and creative thinkers to invent the multitude of concepts and techniques that today make possible for computers to "remediate" other media so well. What were their reasons for doing this? What was their thinking?* In short, why did these people dedicate their careers to inventing the ultimate "remediation machine"?

While media theorists have spent considerable efforts in trying to understand the relationships between digital media and older physical and electronic media in the 1990s and 2000s, the important sources—the writing and projects by Ivan Sutherland, Douglas Engelbart, Ted Nelson, Alan Kay, and other pioneers working in the 1960s and 1970s—remained largely unexamined. This book does not aim to provide a comprehensive intellectual history of the invention of media computing. Thus, I am not going to consider the thinking of all key figures in the history of media computing (to do this right would require more than one book). Rather, my concern is with the present and the future. Specifically, I want to understand some of the dramatic transformations in what media is, what it can do, and how we use it—the transformations that are clearly connected to the shift from previous media technologies to software. Some of these transformations had already taken place in the 1990s but were not much discussed at the time (for instance, the emergence of a new language of moving images and visual design in general). Others have not even been named yet. Still others—such as remix and mashup culture—are being referred to all the time, and yet the analysis of how they were made possible by the evolution of media software has so far not been attempted.

In short, I want to understand what is *"media after software"*—that is, what happened to the techniques, languages, and the concepts of twentieth-century media as a result of their computerization. Or, more precisely, what has happened to media after they have been *software-sized*. (And since in the space of a single book I can only consider some of these techniques, languages and concepts, I will focus on those that, in my opinion, have not been yet discussed by others.)

In this chapter we will take a closer look at one place where the identity of a computer as a "remediation machine" was largely put in place—Alan Kay's Learning Research Group at Xerox PARC, in operation during the 1970s. We can ask two questions: first, what exactly did Kay want to do, and second, how did he and his colleagues go about achieving it? The brief answer—which will be expanded below—is that Kay wanted to turn computers into a "personal dynamic media" which could be used for learning, discovery, and artistic creation. His group achieved this by systematically simulating most existing media within a computer while simultaneously adding many new properties to these media. Kay and his collaborators also developed a new type of programming language that, at least in theory, would allow the users to quickly invent new types of media using the set of general tools already provided for them. All these tools and simulations of already existing media were given a unified user interface designed to activate multiple mentalities and ways of learning—kinesthetic, iconic, and symbolic.

Kay conceived of "personal dynamic media" as a fundamentally new kind of media with a number of historically unprecedented properties such as the ability to hold all the user's information, simulate all types of media within a single machine, and "involve the learner in a two-way conversation."[7] These properties enable new relationships between the user and the media s/he may be creating, editing, or viewing on a computer. And this is essential if we want to understand the relationships between computers and earlier media. Briefly put, while visually, computational media may closely mimic other media, these media now function in different ways.

For instance, consider digital photography, which often imitates traditional photography in appearance. For Bolter and Grusin, this is an example of how digital media 'remediates" its predecessors.

---

[7] Since the work of Kay's group in the 1970s, computer scientists, hackers and designers added many other unique properties—for instance, we can quickly move media around the net and share it with millions of people using Flickr, YouTube, and other sites.

But rather than only paying attention to their appearance, let us think about how digital photographs can function. If a digital photograph is turned into a physical object in the world—an illustration in a magazine, a poster on the wall, a print on a t-shirt—it functions in the same ways as its predecessor (unless it has augmented reality features, like IKEA's 2013 catalog).[8] But if we leave the same photograph inside its native computer environment—which may be a laptop, a network storage system, or any computer-enabled media device such as a cell phone which allows its user to edit this photograph and move it to other devices and the Internet—it can function in ways which, in my view, make it radically different from its traditional equivalent. To use a different term, we can say that a digital photograph offers its users many "affordances" that its non-digital predecessor did not. For example, a digital photograph can be quickly modified in numerous ways and equally quickly combined with other images; instantly moved around the world and shared with other people; and inserted into a multimedia document, or an architectural 3D design. Furthermore, we can automatically (i.e., by running the appropriate algorithms) improve its contrast, make it sharper, and even in some situations remove blur.

Note that only some of these new properties are specific to a particular medium—in our example, a digital photograph, i.e. an array of pixels represented as numbers. Other properties are shared by a larger class of media species—for instance, at the current stage of digital culture, all types of media files can be attached to an email message. Still others are even more general features of a computer environment within the current GUI paradigm as developed forty years ago at PARC: for instance, the fast response of a computer to a user's actions which ensures "no discernible pause between cause and effect."[9] Still others are enabled by network protocols such as TCP/IP that allow all kinds of computers and other devices to be connected to the same network. In summary, we can say that only some of the "new DNA" of a digital photograph is due its particular place of birth, i.e., inside a digital camera. Many others

---

[8] Roberto Baldwin, "Ikea's Augmented Reality Catalog Will Let You Peek Inside Furniture," July 20, 2012, http://www.wired.com/gadgetlab/2012/07/ikeas-augmented-reality-catalog-lets-you-peek-inside-the-malm/

[9] Kay and Goldberg, *Personal Dynamic Media*, p. 394.

are the result of the current paradigm of network computing in general.

Before diving further into Kay's ideas, I should more fully disclose my reasons for focusing on him as opposed to somebody else. The story I will present could also be told differently. It is possible to put Sutherland's work on Sketchpad in the center of computational media history; or Engelbart and his Research Center for Augmenting Human Intellect which throughout the 1960s developed hypertext (independently of Nelson), the mouse, the window, the word processor, mixed text/graphics displays, and a number of other "firsts." Or we can shift focus to the work of the Architecture Machine Group at MIT, which since 1967 was headed by Nicholas Negroponte (in 1985 this group became the MIT Media Lab). We also need to recall that by the time Kay's Learning Research Group at PARC fleshed out the details of GUI and programmed various media editors in Smalltalk (a paint program, an illustration program, an animation program, etc.), artists, filmmakers and architects were already using computers for more than a decade and a number of large-scale exhibitions of computer art were put in major museums around the world such as the Institute of Contemporary Art, London, The Jewish Museum, New York, and Los Angeles County Museum of Art. And certainly, in terms of advancing computer techniques for visual representation enabled by computers, other groups of computer scientists were already ahead. For instance, at the University of Utah, which became the main place for computer graphics research during the first half of the 1970s, scientists were producing 3D computer graphics far superior to the simple images that could be created on computers being built at PARC. Next to the University of Utah, a company called Evans and Sutherland (headed by the same Ivan Sutherland who was also teaching at the University of Utah) was already using 3D graphics for flight simulators—essentially pioneering the type of new media that can be called "navigable 3D virtual space."

While the practical work accomplished at Xerox PARC to establish the computer as a comprehensive media machine is one of my reasons, it is not the only one. The key reason I decided to focus on Kay is his theoretical formulations that place computers in relation to other media and media history. While Vannevar Bush, J. C. R. Licklider and Douglas Engelbart were primary concerned with augmentation of intellectual and in particular scientific work, Kay was equally interested in computers as "a medium of expression through drawing, painting, animating pictures, and composing and generating music."[10] Therefore if we really want to understand how and why computers were redefined as a culture machine, and how the new computational media is different from earlier physical and electronic media, I think that Kay provides us with the best theoretical perspective.

"Simulation is the central notion of the Dynabook"

While Alan Kay articulated his ideas in a number of articles and talks, his 1977 article co-authored with one of his main PARC collaborators, computer scientist Adele Goldberg, is a particularly useful resource if we want to understand contemporary computational media. In this article Kay and Goldberg describe the vision of the Learning Research Group at PARC in the following way: to create "*a personal dynamic medium* the size of a notebook (the Dynabook) which could be owned by everyone and could have the power to handle virtually all of its owner's information-related needs."[11] (The actual Alto computer built at Xerox PARC was the size of later PCs; the article strategically refers to it as "interim dynabook.") Kay and Goldberg ask the readers to imagine that this device "had enough power to outrace your senses of sight and hearing, enough capacity to store for later retrieval thousands of page-equivalents of reference materials, poems, letters, recipes, records, drawings, animations, musical scores, waveforms, dynamic simulations and anything else you would like to remember and change."[12]

In my view, "all" in the first statement is important: it means that the Dynabook—or computational media environment in general, regardless of the size of a form of device in which it

---

[10] *Ibid.*, p. 393.

[11] *Ibid.*, p. 393. The emphasis in this and all following quotes from this article is mine—L. M.

[12] *Ibid.*, p. 394.

is implemented—should support viewing, creating and editing all possible media traditionally used for human expression and communication. Accordingly, while separate programs to create works in different media were already in existence, Kay's group for the first time implemented them all together within a single machine. In other words, Kay's paradigm was not to simply create a new type of computer-based media that would co-exist with other physical media. Rather, the goal was to establish a computer as an umbrella, a platform for *all* existing expressive artistic media. (At the end of the article Kay and Goldberg give a name for this platform, calling it a "metamedium.") This paradigm changes our understanding of what media is. From Gotthold Ephraim Lessing's *Laocoon; or, On the Limits of Painting and Poetry* (1766) to Nelson Goodman's *Languages of Art* (1968), the modern discourse about media depends on the assumption that different mediums have distinct properties and in fact should be understood in opposition to each other. Putting all mediums within a single computer environment does not necessarily erase all differences in what various mediums can represent and how they are perceived—but it does bring them closer to each other in a number of ways. Some of these new connections were already apparent to Kay and his colleagues; others became visible only decades later when the new logic of media set in place at PARC unfolded more fully; some may still not be visible to us today because they have not been given practical realization. One obvious example of such connections is the emergence of multimedia as a standard form of communication: web pages, PowerPoint presentations, multimedia artwork, mobile multimedia messages, media blogs, and other communication forms which combine multiple mediums. Another is the adoption of common interface conventions and tools which we use in working with different types of media regardless of their origin: for instance, a virtual camera, a magnifying lens, and of course the omnipresent copy, cut and paste commands. Yet another is the ability to map one media into another using appropriate software—images into sound, sound into images, quantitative data into a 3D shape or sound, etc.—used widely today in such areas as DJ/VJ/live cinema performances and information visualization. All in all, it is as though different media are actively trying to reach towards each other, exchanging properties and letting each other borrow their unique features. (This situation is the direct opposite

"Kids learning to use the interim Dynabook." (The original caption from the article.)

"The interim Dynabook system consists of processor, disk drive, display, keyboard, and pointing devices." (The original caption from the article.)

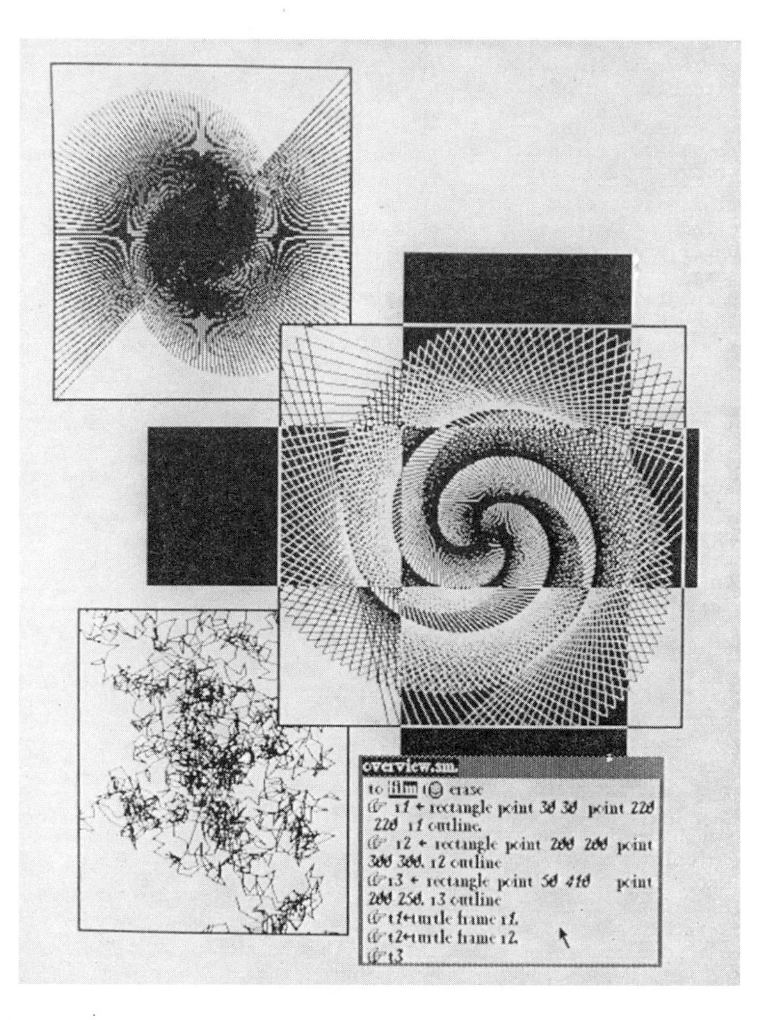

The Alto Screen showing windows with graphics drawn using commands in Smalltalk programming language.

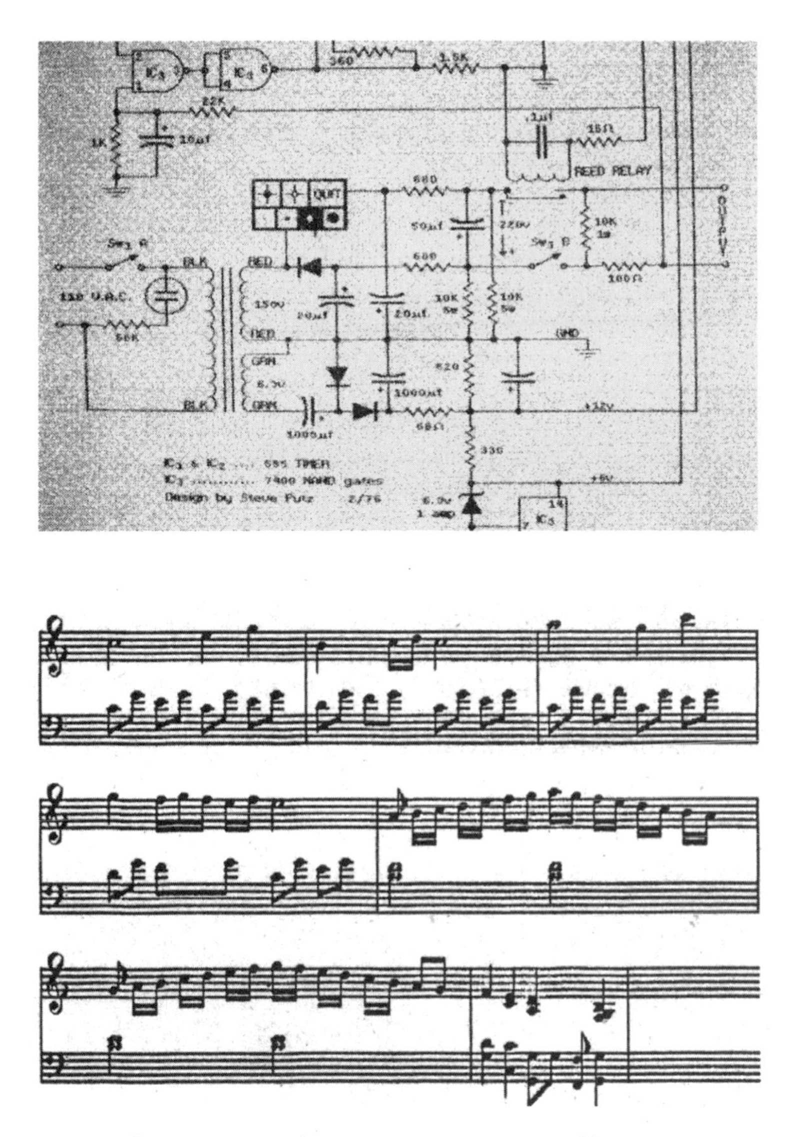

Top: "An electronic circuit layout system programmed by a 15-year-old student" Bottom: "Data for this score was captured on a musical keyboard. A program then converts the data to standard musical notation." (The original captions from the article.)

of the modernist media paradigm of the early twentieth century, which was focused on discovering a unique language for each artistic medium.)

Alan Turing theoretically defined a computer as a machine that can simulate a very large class of other machines, and it is this simulation ability that is largely responsible for the proliferation of computers in modern society. But as I have already mentioned, neither he nor other theorists and inventors of digital computers explicitly considered that this simulation could also include media. It was only Kay and his generation that extended the idea of simulation to media—thus turning Universal Turing Machine into a *Universal Media Machine*, so to speak.

Accordingly, Kay and Goldberg write: "In a very real sense, simulation is the central notion of the Dynabook."[13] When we use computers to simulate some process in the real world—the behavior of a weather system, the processing of information in the brain, the deformation of a car in a crash—our concern is to correctly model the necessary features of this process or system. We want to be able to test how our model would behave in different conditions with different data, and the last thing we want to do is for computers to introduce some new properties into the model that we ourselves did not specify. In short, when we use computers as a general-purpose medium for simulation, we want this medium to be completely "transparent."

But what happens when we simulate different media in a computer? In this case, the appearance of new properties may be welcome as they can extend the expressive and communication potential of these media. Appropriately, when Kay and his colleagues created computer simulations of existing physical media—i.e. the tools for representing, creating, editing, and viewing these media—they "added" many new properties. For instance, in the case of a book, Kay and Goldberg point out "It need not be treated as a simulated paper book since this is *a new medium with new properties*. A dynamic search may be made for a particular context. The non-sequential nature of the file medium and the use of dynamic manipulation allow a story to have many accessible points of view."[14] Kay and his colleagues also added various other

[13] *Ibid.*, p. 399.
[14] *Ibid.*, p. 395.

properties to the computer simulation of paper documents. As Kay has referred to this in another article, his idea was not to simply imitate paper but rather to create "magical paper."[15] For instance, the PARC team gave users the ability to modify the fonts in a document and create new fonts. They also implemented another important idea that had already been developed by Douglas Engelbart's team in the 1960s: the ability to create different views of the same structure (I will discuss this in more detail below). And both Engelbart and Ted Nelson had already "added" something else: the ability to connect different documents or different parts of the same document through hyperlinking—i.e. what we now know as hypertext and hypermedia. Engelbart's group also developed the ability for multiple users to collaborate on the same document. This list goes on and on: e-mail in 1965, newsgroups in 1979, World Wide Web in 1990, etc.

Each of these new properties had far-reaching consequences. Take Search, for instance. Although the ability to search through a page-long text document does not sound like a very radical innovation, as the document gets longer this ability becomes more and more important. It becomes absolutely crucial if we have a very large collection of documents—such as all the web pages on the Web. Although current search engines are far from being perfect and new technologies will continue to evolve, imagine how different the culture of the Web would be without them.

Or take the capacity to collaborate on the same document(s) by a number of users connected to the same network. While it was already widely used by companies in the 1980s and 1990s, it was not until the early 2000s that the wider public saw the real cultural potential of this "addition" to print media. By harvesting the small amounts of labor and expertise contributed by a large number of volunteers, social software projects—most famously, Wikipedia—created vast and dynamically updatable pools of knowledge which would be impossible to create in traditional ways. (In a less visible way, every time we do a search on the Web and then click on some of the results, we also contribute to a knowledge-set used by everybody else. In deciding in which sequence to present the results of a particular search, Google's algorithms take into account which

[15] Alan Kay, "User Interface: A Personal View," p. 199.

among the results of previous searches for the same words people found most useful.)

Studying the writings and public presentations of the people who invented interactive media computing—Sutherland, Engelbart, Nelson, Negroponte, Kay, and others—makes it clear that they did not produce the new properties of computational media as an afterthought. On the contrary, they knew that they were turning physical media into new media. In 1968 Engelbart gave his famous demo at the Fall Joint Computer Conference in San Francisco before a few thousand people that included computer scientists, IBM engineers, people from other companies involved in computers, and funding officers from various government agencies.[16] Although Engelbart had only ninety minutes, he had a lot to show. Over the few previous years, his team at The Research Center for Augmenting Human Intellect had essentially developed the modern *office* environment as it exists today (not be confused with the modern *media design* environment which was developed later at PARC). Their NLS computer system included word processing with outlining features, documents connected through hypertext, online collaboration (two people at remote locations working on the same document in real-time), online user manuals, online project planning systems, and other elements of what is now called "computer-supported collaborative work." The team also developed the key elements of modern user interface that were later refined at PARC: a mouse and multiple windows.

Paying attention to the sequence of the demo reveals that while Engelbart had to make sure that his audience would be able to relate the new computer system to what they already knew and used, his focus was on new features of simulated media never before available previously.[17] Engelbart devotes the first segment of the demo to word processing, but as soon as he briefly demonstrated text entry, cut, paste, insert, naming and saving files—in other words, the set of tools which make a computer into a more versatile typewriter—

---

[16] M. Mitchell Waldrop, *The Dream Machine: J. C. R. Licklider and the Revolution That Made Computing Personal* (Viking, 2001), p. 287.

[17] Complete video of Engelbardt's 1968 demo is available at http://sloan.stanford.edu/MouseSite/1968Demo.html. For the detailed descriptions of NLS functions, see Augmentation Research Center, "NLS User Training Guide," Stanford Research Institute: Menlo Park, California), 1997, http://bitsavers.org/pdf/sri/arc/NLS_User_Training_Guide_Apr77.pdf

he then goes on to show in more depth the features of his system which no writing medium had before: "view control." As Engelbart points out, the new writing medium could switch at the user's wish between *many different views of the same information*. A text file could be sorted in different ways. It could also be organized as a hierarchy with a number of levels, as in outline processors or outlining mode of contemporary word processors such as Microsoft Word. For example, a list of items can be organized by categories and individual categories can be collapsed and expanded.

Engelbart next shows another example of view control, which today, forty-five years after his demo, is still not available in popular document management software. He makes a long "to do" list and organizes it by locations. He then instructs the computer to display these locations as a visual graph (a set of points connected by lines.) In front of our eyes, representation in one medium changes into another medium—text becomes a graph. But this is not all. The user can control this graph to display different amounts of information—something that no image in physical media can do. As Engelbart clicks on different points in a graph corresponding to particular locations, the graph shows the appropriate part of his "to do" list. (This ability to interactively change how much and what information an image shows is particularly important in today's information visualization applications.)

Next Engelbart presents "a chain of views" which he prepared beforehand. He switches between these views using "links" which may look like hyperlinks the way they exist on the Web today—but they actually have a different function. Instead of creating a path between many different documents *à la* Vannevar Bush's Memex (often seen as the precursor to modern hypertext), Engelbart is using links as a method for switching between different views of a single document organized hierarchically. He brings a line of words displayed in the upper part of the screen; when he clicks on these words, more detailed information is displayed in the lower part of the screen. This information can in turn contain links to other views that show even more detail.[18]

---

[18] For the detailed descriptions of these and other capabilities of NLS, see Augmentation Research Center, "NLS User Training Guide," Stanford Research Institute: Menlo Park, California), 1997, http://bitsavers.org/pdf/sri/arc/NLS_User_Training_Guide_Apr77.pdf

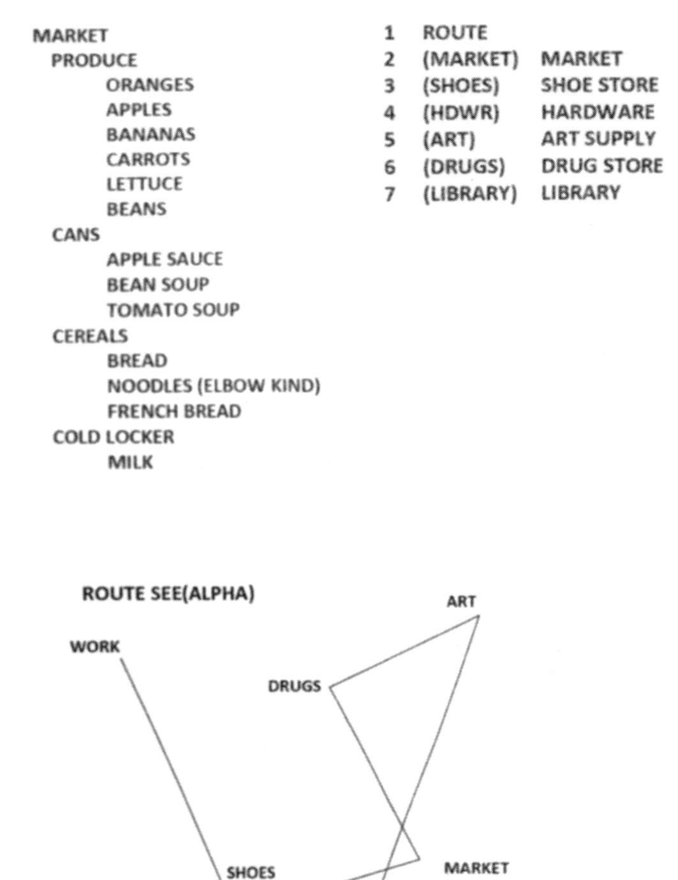

Examples of "view control" as implemented in NLS. Top left: a hierarchical view of a shopping list. Top right: a collapsed view sorted by location. Bottom: a graph view showing the sequence of locations. (Text and graphics were traced from the original video of Engelbart's 1968 demo.)

Rather than using links to drift through the textual universe associatively and "horizontally," we move "vertically" between more general and more detailed information. Appropriately, in Engelbart's paradigm, we are not "navigating"—we are "switching views." We can create many different views of the same information and switch between these views in different ways. And this is what Engelbart systematically explains in this first part of his demo. He demonstrates that you can change views by issuing commands, by typing numbers that correspond to different parts of a hierarchy, by clicking on parts of a picture, or on links in the text. (In 1967 Ted Nelson articulated and named a similar idea of a type of hypertext, which would allow a reader to "obtain a greater detail on a specific subject." He named it "stretchtext."[19])

Since new media theory and criticism emerged in the early 1990s, endless texts have been written about interactivity, hypertext, virtual reality, cyberspace, cyberculture, cyborgs, and so on. But I have never seen anybody discuss "view control." And yet this is one of the most fundamental and radical new techniques for working with information and media available to us today. It is used daily by each of us numerous times. "View control," i.e. the abilities to switch between many different views and kinds of views of the same information is now implemented in multiple ways not only in OS, word processors and email clients, but also in all "media processors" (i.e. media editing software): AutoCAD, Maya, After Effects, Final Cut, Photoshop, InDesign, and so on. For instance, in the case of 3D software, it can usually display the model in at least half a dozen different ways: in wireframe, fully rendered, etc. In the case of animation and visual effects software, since a typical project may contain dozens of separate objects each having dozens of parameters, it is often displayed in a way similar to how outline processors can show text. In other words, the user can switch between more and less information. You can choose to see only those parameters which you are working on right now. You can also zoom in and out of the composition. When you do this, parts of the composition do not simply get smaller or bigger—they show less or more information automatically. For instance, at a certain scale you may only see the names of different

[19] Ted Nelson, "Stretchtext" (Hypertext Note 8), 1967, http://xanadu.com/XUarchive/htn8.tif

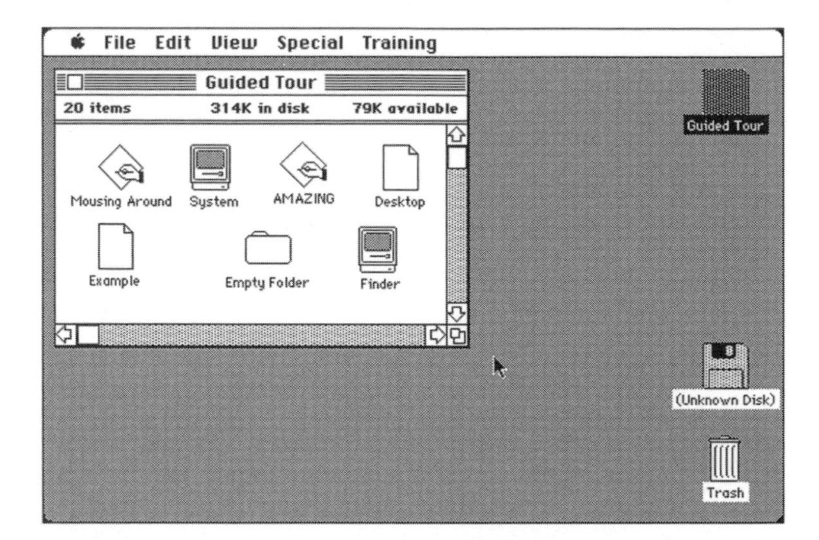

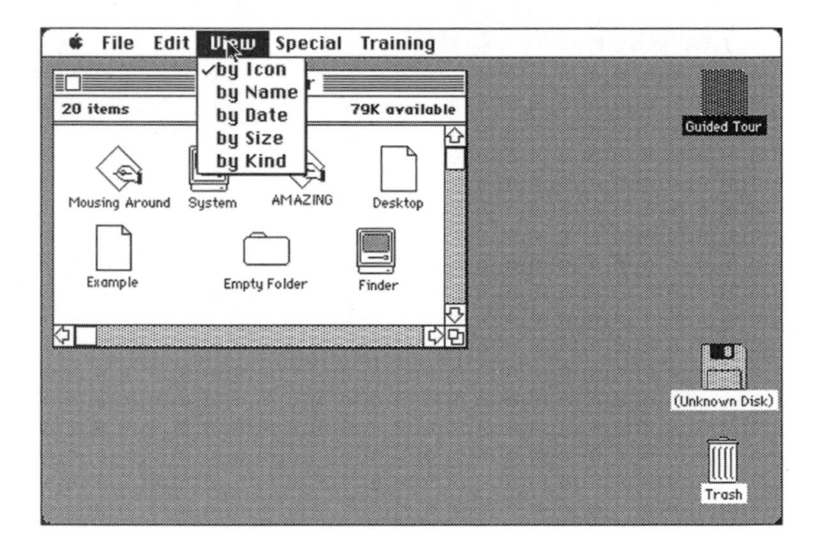

View control as implemented in Macintosh System software, 1984. Top: applications, folders, and files in "Guided Tour" floppy disk. Bottom: View of applications, folders, and files sorted by icon.

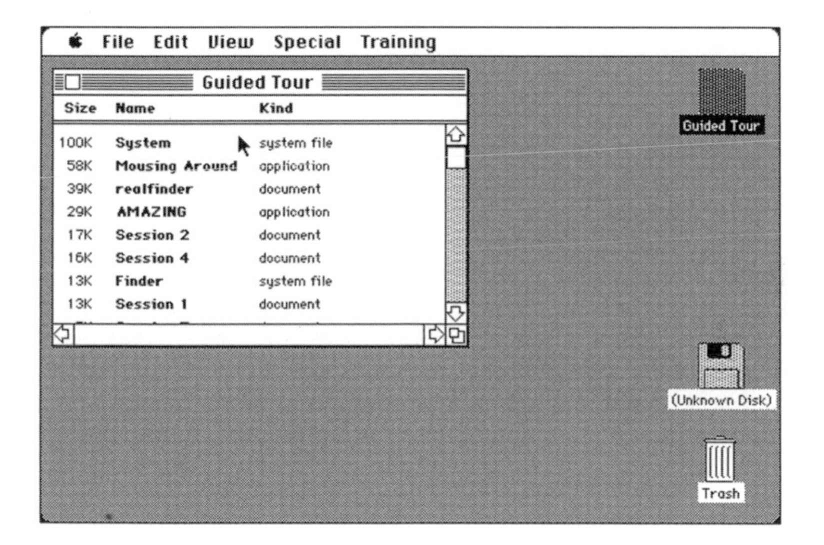

Top: View of applications, folders, and files sorted by date. Bottom: View of applications, folders, and files sorted by size.

parameters; but when you zoom into the display, the program may also display the graphs which indicate how these parameters change over time.

Let us look at another example—Ted Nelson's concept of hypertext that he developed in the early 1960s (independently but parallel to Engelbart).[20] In his 1965 article *A File Structure for the Complex, the Changing, and the Indeterminate*, Nelson discusses the limitations of books and other paper-based systems for organizing information and then introduces his new concept:

> However, with the computer-driven display and mass memory, it has become possible to create *a new, readable medium*, for education and enjoyment, that will let the reader find his level, suit his taste, and find the parts that take on special meaning for him, as instruction and enjoyment.
>
> Let me introduce the word "hypertext" to mean a body of written or pictorial material interconnected in such a complex way that it could not be conveniently presented or represented on paper.[21]

"A new, readable medium"—these words make it clear that Nelson was not simply interested in "patching up" books and other paper documents. Instead, he wanted to create something distinctively new. But was not hypertext as proposed by Nelson simply an extension of older textual practices such as exegesis (extensive interpretations of holy scriptures such as the Bible, Talmud, Qur'ān), annotations, or footnotes? While such historical precedents for hypertext are often proposed, they mistakenly equate Nelson's proposal with a very limited form in which hypertext is experienced by most people today—i.e., the World Wide Web. As Noah Wardrip-Fruin pointed out, "The Web implemented only

---

[20] Douglas C. Engelbart, *Augmenting Human Intellect: A Conceptual Framework* (Stanford Research Institute, 1962), http://www.dougengelbart.org/pubs/augment-3906.html. Although the implementation of hypertext in Engelbart's NLS was much more limited than Nelson's concept of hypertext, looking at Engelbart's discussion in *Augmenting Human Intellect* shows that his ideas for new systems for organizing information were at least as rich as Nelson's.

[21] Theodor H. Nelson, "A File Structure for the Complex, the Changing, and the Indeterminate" (1965), in *New Media Reader*, p. 144.

one of many types of structures proposed by Nelson already in 1965—'chunk style' hypertext—static links that allow the user to jump from page to page."[22]

Following the Web implementation, most people today think of hypertext as a body of text connected through one-directional links. However, the terms "links" does not even appear in Nelson's original definition of hypertext. Instead, Nelson talks about new complex interconnectivity without specifying any particular mechanisms that can be employed to achieve it. A particular system proposed in Nelson's 1965 article is one way to implement such a vision, but as his definition implicitly suggests, many others are also possible.

"What kind of structures are possible in hypertext?" asks Nelson in a research note from 1967. He answers his own question in a short but very suggestive manner: "Any."[23] Nelson goes on to explain: "Ordinary text may be regarded as a special case—the simple and familiar case—of hypertext, just as three-dimensional space and the ordinary cube are the simple and familiar special cases of hyperspace and hypercube."[24] (In 2007 Nelson re-stated this idea in the following way: " 'Hypertext'—a word I coined long ago—is not technology but potentially the fullest generalization of documents and literature."[25])

If "hypertext" does not simply mean "links," it also does not only mean "text." Although in its later popular use the word "hypertext" came to refer to linked text, as one can see from the quote above, Nelson included "pictures" in his definition of hypertext.[26] And in the following paragraph, he introduces the terms *hyperfilm* and *hypermedia*:

---

[22] Noah Wardrip-Fruin, introduction to Theodor H. Nelson, "A File Structure for the Complex, the Changing, and the Indeterminate" (1965), in *New Media Reader*, p. 133.

[23] Ted Nelson, "Brief Words on the Hypertext" (Hypertext Note 1), 1967, http://xanadu.com/XUarchive/htn1.tif

[24] *Ibid.*

[25] Ted Nelson, http://transliterature.org/ (version TransHum-D23, 07.06.17).

[26] In his presentation at the 2004 Digital Retroaction symposium Noah Wardrip-Fruin stressed that Nelson's vision included hypermedia and not only hypertext. Noah Wardrip-Fruin, presentation at Digital Retroaction: a Research Symposium, UC Santa Barbara, September 17–19, 2005, http://dc-mrg.english.ucsb.edu/conference/D_Retro/conference.html

> Films, sound recordings, and video recordings are also linear strings, basically for mechanical reasons. But these, too, can now be arranged as non-linear systems – for instance, lattices – for educational purposes, or for display with different emphasis… The hyperfilm – a browsable or vari-sequenced movie – is only one of the possible hypermedia that require our attention."[27]

Where is hyperfilm today, almost 50 years after Nelson articulated this concept? If we understand hyperfilm in the same limited sense as hypertext is understood today—shots connected through links which a user can click on—it would seems that hyperfilm never fully took off. A number of early pioneering projects—*Aspen Movie Map* (Architecture Machine Group, 1978–9), *Earl King* and *Sonata* (Grahame Weinbren, 1983–5; 1991–3), CD-ROMs by Bob Stein's Voyager Company, and *Wax: Or the Discovery of Television Among the Bees* (David Blair, 1993)—have not been followed up. Similarly, interactive movies and FMV-games created by the video game industry in the first half of the 1990s soon fell out of favor, replaced by 3D games (which offered more interactivity). But if instead we think of hyperfilm in a broader sense, as it was conceived by Nelson—any interactive structure for connecting video or film elements, with a traditional film being a special case—we realize that hyperfilm is much more common today than it may appear. Numerous interactive Flash and HTML5 sites which use video, video clips with markers which allow a user jump to a particular point in a video (for instance, see the videos on TED.com[28]), and database cinema[29] are just some of the examples of hyperfilm today.

Decades before hypertext and hypermedia became the common ways for interacting with information, Nelson understood well what these ideas meant for our well-established cultural practices and concepts. The announcement for his January 5, 1965 lecture at Vassar College talks about this in terms that are even more relevant today than they were then: "The philosophical consequences of all this are very grave. Our concepts of 'reading', 'writing', and 'book' fall apart, and we are challenged to design 'hyperfiles' and write

[27] Nelson, *A File Structure*, p. 144.
[28] www.ted.com (March 8, 2008).
[29] See http://softcinema.net/form.htm

'hypertext' that may have more teaching power than anything that could ever be printed on paper."[30]

These statements align Nelson's thinking and work with artists and theorists who similarly wanted to destabilize the conventions of cultural communication. Digital media scholars extensively discussed parallels between Nelson and French theorists writing during the 1960s—Roland Barthes, Michel Foucault and Jacque Derrida.[31] Others have pointed out close parallels between the thinking of Nelson and literary experiments taking place around the same time, such as works by Oulipo.[32] (We can also note the connection between Nelson's hypertext and the non-linear structure of the films of French filmmakers who set out to question the classical narrative style: *Hiroshima Mon Amour, Last Year at Marienbad*, *Breathless* and others).

How far shall we take these parallels? In 1987 Jay Bolter and Michael Joyce wrote that hypertext could be seen as "a continuation of the modern 'tradition' of experimental literature in print" which includes "modernism, futurism, Dada surrealism, lettrism, the nouveau roman, concrete poetry."[33] Refuting their claim, Espen J. Aarseth has argued that hypertext is not a modernist structure *per se*, although it can support modernist poetics if the author desires this.[34] Who is right? Since this book argues that cultural software turned media into metamedia—a fundamentally new semiotic and technological system which includes most previous media techniques and aesthetics as its elements—I also think that hypertext is actually quite different from modernist literary tradition. I agree with Aarseth that hypertext is indeed much more general than any particular poetics such as modernist ones.

---

[30] Announcement of Ted Nelson's lecture at Vassar College, January 5, 1965, http://xanadu.com/XUarchive/ccnwwt65.tif

[31] George Landow, ed., *Hypertext: The Convergence of Contemporary Critical Theory and Technology* (The Johns Hopkins University Press, 1991); Jay Bolter, *The writing space: the computer, hypertext, and the history of writing* (Hillsdale, NJ: L. Erlbaum Associates, 1991).

[32] Randall Packer and Ken Jordan, *Multimedia: From Wagner to Virtual Reality* (W. W. Norton & Company, 2001); Noah Wardrip-Fruin and Nick Monford, *New Media Reader* (The MIT Press, 2003).

[33] Quoted in Espen J. Aarseth, *Cybertext: Perspectives on Ergodic Literature* (The Johns Hopkins University Press, 1997), p. 89.

[34] Espen J. Aarseth, *Cybertext*, 89–90.

Indeed, already in 1967 Nelson said that hypertext could support any structure of information including that of traditional texts—and presumably, this also includes different modernist poetics. (Importantly, this statement is echoed in Kay and Goldberg's definition of the computer as a "metamedium" whose content is "a wide range of already-existing and not-yet-invented media.")

What about the scholars who see the strong connections between the thinking of Nelson and modernism? Although Nelson says that hypertext can support any information structure and that this information does not need to be limited to text, his examples and his style of writing show an unmistakable aesthetic sensibility—that of literary modernism. He clearly dislikes "ordinary text." The emphasis on *complexity and interconnectivity* and on *breaking up conventional units for organizing information* such as a page clearly aligns Nelson's proposal for hypertext with the early twentieth-century experimental literature—the inventions of Virginia Woolf, James Joyce, the Surrealists, etc. This connection to literature is not accidental since Nelson's original motivation for his research that led to hypertext was to create a system for handling both the notes for literary manuscripts and those manuscripts themselves. Nelson also already knew about the writings of William Burroughs. The very title of the article—*A File Structure for the Complex, the Changing, and the Indeterminate*—would make the perfect title for an early twentieth-century avant-garde manifesto, as long as we substitute "file structure" with some "ism."

Nelson's modernist sensibility also shows itself in his thinking about new mediums that can be established with the help of a computer. However, his work should not be seen as a simple continuation of modernist tradition. Rather, both his and Kay's research represent the next stage of the avant-garde project. The early twentieth-century avant-garde artists were primarily interested in questioning conventions of established media such as photography, print, graphic design, cinema, and architecture. Thus, no matter how unconventional the paintings that came out from Futurism, Orphism, Suprematism or De Stijl were, their manifestos were still talking about them as paintings—rather than as a new media. In contrast, Nelson and Kay explicitly write about creating new media, not only changing the existing ones. Nelson: "With the computer-driven display and mass memory, it has become possible to create a new, readable medium." Kay and Goldberg: "It [computer text] need not be treated as a simulated paper book since this is a new medium with new properties."

Another key difference between how modernist artists and pioneers of cultural software approached the job of inventing new media and extending existing ones is captured by the title of Nelson's article I have been already quoting above: "*A File Structure for the Complex, the Changing, and the Indeterminate*." Instead of a particular modernist "ism," we get a file structure. Cubism, Expressionism, Futurism, Orphism, Suprematism, and Surrealism proposed new distinct systems for organizing information, with each system fighting all others for the dominance in the cultural memesphere. In contrast, Bush, Licklider, Nelson, Engelbart, Kay, Negroponte, and their colleagues created meta-systems that can support many kinds of information structures. Kay called such a system "a first metamedium," Nelson referred to it as hypertext and hypermedia, Engelbart wrote about "automated external symbol manipulation" and "bootstrapping,"—but behind the differences in their visions lay the similar understanding of the radically new potential offered by computers for information manipulation. The prefixes "meta-" and "hyper-" used by Kay and Nelson were the appropriate characterizations for a system which was more than another new medium that could remediate other media in its particular ways. Instead, the new system would be capable of simulating all these media with all their remediation strategies—as well as supporting development of what Kay and Goldberg referred to as new "not-yet-invented media." And of course, this was not all. Equally important was the role of interactivity. The new meta-systems proposed by Nelson, Kay and others were to be used interactively to support the processes of thinking, discovery, decision making, and creative expression. In contrast, the aesthetics created by modernist movements could be understood as "information formatting" systems—to be used for selecting and organizing information into fixed presentations that are then distributed to the users, not unlike PowerPoint slides. Finally, at least in Kay's and Nelson's vision, the task of defining new information structures and media manipulation techniques—and, in fact, new media as a whole—was given to the user, rather than being the sole province of the designers. This decision had far-reaching consequences for shaping contemporary culture. Once computers and programming were democratized enough, many creative people started to focus on creating these new structures and techniques rather than using the existing ones to make "content." Since the end of 2000, extending the computer metamedium by writing new software, plugins, programming libraries and other tools became the new cutting-edge type of cultural activity – giving a new meaning to McLuhan's famous formula "the medium is the message."

Today a typical article in computer science or information science will not be talking about inventing a "new medium" as a justification for research. Instead, it is likely to refer to previous work in some field or sub-field of computer science such as "knowledge discovery," "data mining," "semantic web," etc. It can also refer to existing social and cultural practices and industries—for instance, "e-learning," "video game development," "collaborative tagging," or "massively distributed collaboration." In either case, the need for new research is justified by a reference to already established or popular practices—academic paradigms which have been funded, large-scale industries, and mainstream social routines which do not threaten or question the existing social order. This means that practically all of computer science research which deals with media—web technologies, media computing, hypermedia, human-computer interfaces, computer graphics, and so on—is oriented towards "mainstream" media usage.

In other words, either computer scientists are trying to make more efficient the technologies already used in media industries (video games, web search engines, film production, etc.) or they are inventing new technologies that are likely to be used by these industries in the future. The invention of new mediums for its own sake is not something which anybody is likely to pursue, or get funded. From this perspective, the software industry and business in general is often more innovative than academic computer science. For instance, social media applications (Wikipedia, Flickr, YouTube, Facebook, del.icio.us, Digg, etc.) were not invented in the academy; nor were HyperCard, QuickTime, HTML, Photoshop, After Effects, Flash, or Google Earth. This was no different in previous decades. It is, therefore, not accidental that the careers of both Ted Nelson and Alan Kay were spent in the industry and not the academy: Kay worked for and was a fellow at Xerox PARC, Atari, Apple and Hewlett-Packard; Nelson was a consultant and a fellow at Bell Laboratories, Datapoint Corporation, Autodesk; both were also associated with Disney.

Why did Nelson and Kay find more support in industry than in academia for their quest to invent new computer media? And why is the industry (by which I simply mean any entity which creates the products which can be sold in large quantities, or monetized in other ways, regardless of whether this entity is a large multinational company or a small start-up)—more interested in innovative media technologies, applications, and content than computer science? The systematic answer to this question will require its own investigation. Also, what kinds of innovations each modern institution can support changes over time. But here is one brief answer: modern business thrives on creating new markets, new products, and new product categories. Although the actual development of such new markets and products is always risky, it is also very profitable. This was already the case in the previous decades when Nelson and Kay were supported by Xerox, Atari, Apple, Bell Labs, Disney, etc. In the 2000s, following the globalization of the 1990s, all areas of business embraced innovation to an unprecedented degree; this pace quickened around 2005 as companies fully focused on competing for new consumers in China, India, and other "emerging" economies. Around the same time, we saw a similar increase in the number of innovative products in the IT industry: open APIs of leading Web 2.0 sites, daily announcements of new web services, locative media applications, new innovative products such as iPhone, new paradigms in imaging such as HDR and non-destructive editing, the beginnings of a "long tail" for software, open source hardware, and so on.

As we can see from the examples we have analyzed, the aim of the inventors of computational media—Engelbart, Nelson, Kay and the people who worked with them—was not simply to create accurate simulations of physical media. Instead, in every case the goal was to create "a new medium with new properties" which would allow people to communicate, learn, and create in new ways. So while today the content of these new media may often look the same as that of its predecessors, we should not be fooled by this similarity. The newness lies not in the content but in the software tools used to create, edit, view, distribute, and share this content. Therefore, rather than only looking at the "output" of software-based cultural practices, we need to consider software itself—since it allows people to work with media in a number of historically unprecedented ways. So while on the level of appearance computational media indeed often remediate (i.e. represent) previous media, the software environment in which this media "lives" is very different.

Let me add two more examples. One is Ivan Sutherland's *Sketchpad* (1962). Created by Sutherland as a part of his PhD thesis at MIT, Sketchpad deeply influenced all subsequent work in computational media (including that of Kay) not only because it was the first interactive media authoring program but also because it made it clear that computer simulations of physical media can add many exciting new properties to the media being simulated. Sketchpad was the first software that allowed its users to interactively create and modify line drawings. As Noah Wardrip-Fruin pointed out, it "moved beyond paper by allowing the user to work at any of 2000 levels of magnification—enabling the creation of projects that, in physical media, would either be unwieldy large or require detail work at an impractically small size."[35] Sketchpad similarly redefined graphical elements of a design as objects which "can be manipulated, constrained, instantiated, represented ironically, copied, and recursively operated upon, even recursively merged.'[36] For instance, if the designer defined new graphical elements as instances of a master element and later made a change to the master, all these instances would also change automatically.

Another new property, which perhaps demonstrated most dramatically how computer-aided drafting and drawing were different from their physical counterparts, was Sketchpad's use of constraints. In Sutherland's own words, "The major feature which distinguishes a Sketchpad drawing from a paper and pencil drawing is the user's ability to specify to Sketchpad mathematical conditions on already drawn parts of his drawing which will be automatically satisfied by the computer to make the drawing take the exact shape desired."[37] For instance, if a user drew a few lines, and then gave the appropriate command, Sketchpad automatically

---

[35] Noah Wardrip-Fruin, introduction to "Sketchpad. A Man-Machine Graphical Communication System," in *New Media Reader*, 1963, p. 109.

[36] *Ibid.*

[37] Ivan Sutherland, "Sketchpad. A Man-Machine Graphical Communication System," *Proceedings of the AFIPS Spring Joint Computer Conference*, Detroit,

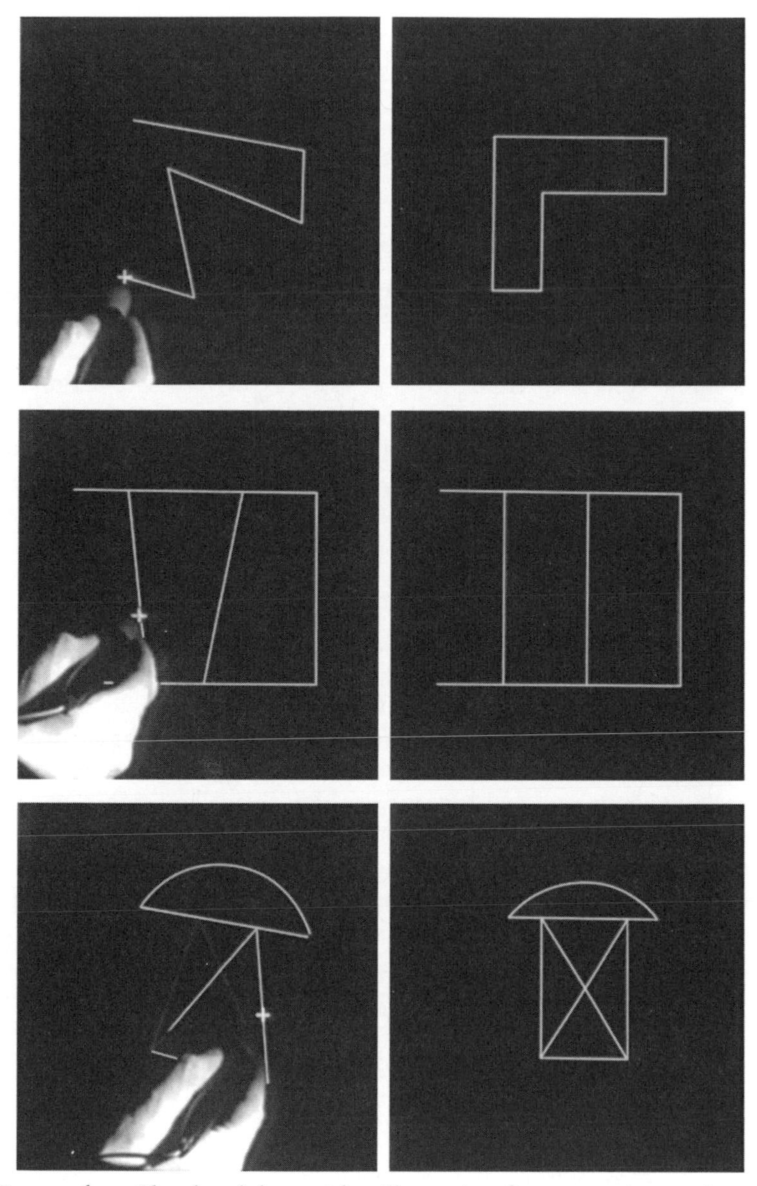

Frames from Sketchpad demo video illustrating the program's use of constraints. Left column: a user selects parts of a drawing. Right column: Sketchpad automatically adjusts the drawing. (The captured frames were edited in Photoshop to show the Sketchpad screen more clearly.)

moved these lines until they were parallel to each other. If a user gave a different command and selected a particular line, Sketchpad moved the lines in such a way so they would parallel to each other and perpendicular to the selected line.

Although we have not exhausted the list of new properties that Sutherland built into Sketchpad, it should be clear that this first interactive graphical editor was not only simulating existing media. Appropriately, Sutherland's 1963 paper on Sketchpad repeatedly emphasizes the new graphical capacities of his system, marveling how it opens new fields of "graphical manipulation that has never been available before."[38] The very title given by Sutherland to his PhD thesis foregrounds the novelty of his work: *Sketchpad: A man-machine graphical communication system*. Rather than conceiving of Sketchpad as simply another medium, Sutherland presents it as something else—a communication system between two entities: a human and an intelligent machine. Kay and Goldberg later also foregrounded this communication dimension, referring to it as "a two-way conversation" and calling the new "metamedium" "active."[39] (We can also think of Sketchpad as a practical demonstration of the idea of "man-machine symbiosis" by J. C. R. Licklider applied to image making and design.[40])

My last example comes from the software development that at first sight may appear to contradict my argument: paint software. Surely, the applications which simulate in detail the range of effects made possible with various physical brushes, paint knives, canvases, and papers are driven by the desire to recreate the experience of working within an existing medium rather than the desire to create a new one? Wrong. In 1997 an important computer graphics pioneer Alvy Ray Smith wrote a memo titled *Digital Paint Systems: Historical Overview*.[41] In this text Smith (who himself had a background in art) makes an important distinction between

---

Michigan, May 21–3, 1963, pp. 329–46; in *New Media Reader*, Noah Wardrip-Fruin and Nick Montfort (eds).

[38] *Ibid*., p. 123.

[39] Kay and Goldberg, "Personal Dynamic Media," 394.

[40] J. C. R. Licklider, "Man-Machine Symbiosis," *IRE Transactions on Human Factors in Electronics,* vol. HFE-1, March 1960, pp. 4–11, in *New Media Reader*, eds. Noah Wardrip-Fruin and Nick Montfort.

[41] Alvy Ray Smith, *Digital Paint Systems: Historical Overview* (Microsoft Technical Memo 14, May 30, 1997). http://alvyray.com/

digital paint programs and *digital paint systems*. In his definition, "A digital paint *program* does essentially no more than implement a digital simulation of classic painting with a brush on a canvas. A digital paint *system* will take the notion much farther, using the "simulation of painting" as a familiar metaphor to seduce the artist into the new digital, and perhaps forbidding, domain." (Emphasis in the original). According to Smith's history, most commercial painting applications, including Photoshop, fall into the paint system category. His genealogy of paint systems begins with Richard Shoup's SuperPaint, developed at Xerox PARC in 1972–3.[42] While SuperPaint allowed the user to paint with a variety of brushes in different colors, it also included many techniques not possible with traditional painting or drawing tools. For instance, as described by Shoup in one of his articles on SuperPaint, "Objects or areas in the picture may be scaled up or down in size, moved, copied, overlaid, combined or changed in color, and saved on disk for future use or erased."[43]

Most important, however, was the ability to grab frames from video. Once loaded into the system, such a frame could be treated as any other image—that is, an artist could use all of SuperPaint's drawing and manipulation tools, add text, combine it with other images, etc. The system could also translate what appeared on its screen back into a video signal. Accordingly, Shoup is clear that his system was much more than a way to draw and paint with a computer. In a 1979 article, he refers to SuperPaint as a new "videographic medium."[44] In another article published a year later, he refines this claim: "From a larger perspective, we realized that the development of SuperPaint signaled the beginning of the synergy of two of the most powerful and pervasive technologies ever invented: digital computing and video or television."[45]

This statement is amazingly perceptive. When Shoup was writing this in 1980, computer graphics were used in television

---

[42] Richard Shoup, "SuperPaint: An Early Frame Buffer Graphics Systems," IEEE Annals of the History of Computing 23, issue 2 (April–June 2001), p. 32–7, http://www.rgshoup.com/prof/SuperPaint/Annals_final.pdf; Richard Shoup, "SuperPaint...The Digital Animator," *Datamation* (1979), http://www.rgshoup.com/prof/SuperPaint/Datamation.pdf.

[43] Shoup, "SuperPaint...The Digital Animator," p. 152.

[44] *Ibid.*, p. 156.

[45] Shoup, "SuperPaint: An Early Frame Buffer Graphics System," p. 32.

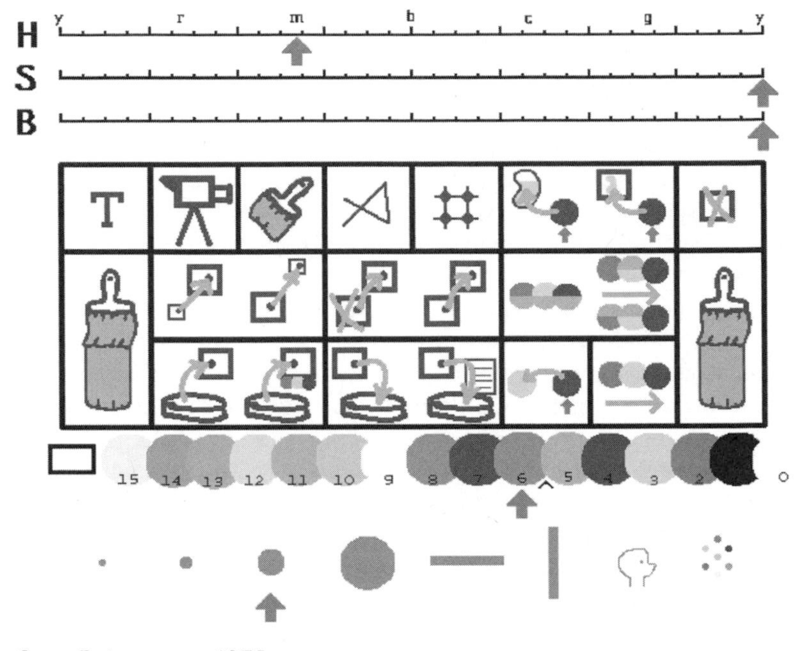

SuperPaint menu, 1975.

broadcasts just a handful of times. And while in the next decade their use became more common, only in the middle of the 1990s did the synergy Shoup predicted truly became visible. As we will see in the chapter on After Effects below, the result was a dramatic reconfiguration not just of the visual languages of television but of all visual techniques invented by humans up to that point. In other words, what began as a new "videographic medium" in 1973 had eventually changed all visual media.

But even if we forget about SuperPaint's revolutionary ability to combine graphics and video, and discount its new tools such resizing, moving, copying, etc., we are still dealing with *a new creative medium* (Smith's term). As Smith pointed out, this medium is the *digital frame buffer*,[46] a special kind of computer memory

[46] Alvy Ray Smith, "Digital Paint Systems: An Anecdotal and Historical Overview," *IEEE Annals of the History of Computing*. 2011, http://accad.osu.edu/~waynec/history/PDFs/paint.pdf

designed to hold images represented as an array of pixels (today a more common name is *graphics card*). An artist using a paint system is modifying pixel values in a frame buffer—regardless of what particular operation or tool s/he is employing at the moment. This opens up a door to all kinds of new image creation and modification operations, which follow different logic than physical painting. The telling examples of this can be found in a paint system called Paint developed by Smith in 1975–6. In Smith's own words, "Instead of just simulating painting a stroke of constant color, I extended the notion to mean 'perform any image manipulation you want under the pixels of the paintbrush.'"[47] Beginning with this conceptual generalization, Smith added a number of effects which still used a paintbrush tool but actually no longer referred to painting in a physical world. For instance, in Paint "any image of any shape could be used as a brush." In another example, Smith added "'not paint' that reversed the color of every pixel under the paintbrush to its color complement." He also defined 'smear paint' that averaged the colors in the neighborhood of each pixel under the brush and wrote the result back into the pixel." And so on. Thus, the instances where the paintbrush tool behaved more like a real physical paintbrush were just particular cases of a much larger universe of new behaviors made possible in a new medium.

The permanent extendibility

As we saw, Sutherland, Nelson, Engelbart, Kay, and other pioneers of computational media have added many previously non existent properties to media that they have simulated in a computer. The subsequent generations of computer scientists, hackers, and designers added many more properties—but this process is far from finished. And there is no logical or material reason why it will ever be finished. It is the "nature" of computational media that it is open-ended and that new techniques are continuously being invented.

To add new properties to physical media requires modifying its physical substance. But since computational media exists as

[47] *Ibid.*, p. 18.

software, we can add new properties or even invent new types of media by simply changing existing or writing new software. Or by adding plug-ins and extensions, as programmers have been doing it with Photoshop and Firefox, respectively. Or by putting existing software together. (For instance, starting in 2006, thousands of people extended the capacities of mapping media by creating software mashups which combine the services and data provided by Goggle Maps, Flickr, Amazon, other sites, and media uploaded by users.)

In short, *"new media" is "new" because new properties (i.e., new software techniques) can always be easily added to it.* Put differently, in industrial (i.e. mass-produced) media technologies, "hardware" and "software" were one and the same thing. For example, the book pages were bound in a particular way that fixed the order of pages. The reader could not change this order nor the level of detail being displayed *à la* Engelbart's "view control." Similarly, the film projector combined hardware and what we now call a "media player" software into a single machine. In the same way, the controls built into a twentieth-century mass-produced camera could not be modified at the user's will. And although today the users of a digital camera similarly cannot easily modify the hardware of their camera, as soon as they transfer the pictures into a computer they have access to endless number of controls and options for modifying their pictures via software.

In the nineteenth and twentieth centuries the normally rigid industrial media was fluid in two situations. First, when a new media was being first developed: for instance, the invention of photography in the 1820s–1840s. Second, when artists would systematically experiment with and "open up" already industrialized media—such as the experiments with film and video during the 1960s that came to be called "Expanded Cinema."

What used to be separate moments of experimentations with media during the industrial era became the norm in a software society. In other words, the computer legitimizes experimentation with media. Why is this so? What differentiates a modern digital computer from any other machine—including industrial media machines for capturing and playing media—is separation of hardware and software. It is because an endless number of different programs performing different tasks can be written to run on the same type of machine, that that machine—i.e. a digital computer—is used so widely today. Consequently, the constant invention of new (and modification of existing) media software, is simply one example of this general principle. In its very structure computational media is "avant-garde" since it is constantly being extended and thus redefined.

If in modern culture "experimental" and "avant-garde" were opposed to normalized and stable, this opposition largely disappears in software culture. And the role of the media avant-garde is performed no longer by individual artists in their studios but by a variety of players, from very big to very small—from companies such as Microsoft, Adobe, and Apple to independent programmers, hackers, and designers.

But this process of continual invention of new algorithms does not just move in any direction. If we look at contemporary media software—CAD, computer drawing and painting, image editing, word processors—we will see that most of their fundamental principles were already developed by the generation of Sutherland and Kay. In fact the very first interactive graphical editor—Sketchpad—already contains most of the genes, so to speak, of contemporary graphics applications. As new techniques continue to be invented they are layered over the foundations that were gradually put in place by Sutherland, Engelbart, Kay, and others in the 1960s and 1970s.

Of course we are not dealing here only with the history of ideas. Various social and economic factors—such as the dominance of the media software market by a handful of companies or the wide adoption of particular file formats — also constrain possible directions of software evolution. Put differently, today software development is an industry and as such it is constantly balancing between stability and innovation, standardization and exploration of new possibilities. But it is not just any industry. New programs can be written and existing programs can be extended and modified (if the source code is available) by anybody who has programming skills and access to a computer, a programming language and a compiler. In other words, today software is fundamentally malleable in a way that twentieth-century industrially produced objects were not. (The emergence of consumer 3D printing and the "open hardware" movement promise to bring such flexibility to physical objects as well, but it will be a while before you can print a whole ready-to-drive-car on your home 3D printer.)

Although Turing and Von Neumann formulated this fundamental extendibility of software in theory, its contemporary practice—hundreds of thousands of people daily involved in extending the capabilities of computational media—is a result of a long historical development. This development took us from the few early room-sized computers, which were not easy to reprogram to a wide availability of cheap computers and programming tools decades later. This democratization of software development was at the core of Kay's vision. Kay was particularly concerned with how to structure programming tools in such a way that would make development of media software possible for ordinary users. For instance, at the end of the 1977 article I have already extensively quoted, he and Goldberg write, "We must also provide enough already-written general tools so that a user need not start from scratch for most things she or he may wish to do."

Comparing the process of continuous media innovation via new software to the history of earlier, pre-computational media reveals a new logic at work. According to a commonplace idea, when a new medium is invented it first closely imitates already existing media, before discovering its own language and aesthetics. Indeed, the first Gutenberg Bible closely imitated the look of the handwritten manuscripts; early films produced in the 1890s and 1900s mimicked the presentational format of theatre by positioning the actors on the invisible shallow stage and having them face the audience. Slowly, printed books developed a different way of presenting information; similarly cinema also developed its own original concept of narrative space. Through repetitive shifts in points of view presented in subsequent shots, the viewers were placed inside this space—thus literally finding themselves inside the story.

Can this logic apply to the history of computer media? As theorized by Turing and Von Neumann, the computer is a general-purpose simulation machine. This is its uniqueness and its difference from all other machines and previous media. This means that the idea that a new medium gradually finds its own language cannot apply to computer media. If this were true it would go against the very definition of a modern digital computer. This theoretical argument is supported by practice. The history of computer media so far has been not about arriving at some standardized language—as, for instance, happened with cinema—but rather about the gradual expansion of uses, techniques, and possibilities. Rather than arriving at a particular language, we are gradually discovering that the computer can speak more and more languages.

If we are to look more closely at the early history of computer media—for instance, the way we have been looking at Kay's ideas and work in this text—we will discover another reason why the idea of a new medium gradually discovering its own language does not apply to computer media. The systematic practical work on making a computer simulate and extend existing media (Sutherland's Sketchpad, the first interactive word processor developed by Engelbart's group, etc.) came after computers had already been put to multiple uses—performing different types of calculations, solving mathematical problems, controlling other machines in real time, running mathematical simulations, simulating some aspects of human intelligence, and so on. (We should also mention the work on SAGE by MIT Lincoln Laboratory which, by the middle of the 1950s, had already established the idea of interactive communication between a human and a computer via a screen with a graphical display and a pointing device. In fact, Sutherland developed Sketchpad on a TX-2, the new version of a larger computer MIT constructed for SAGE.) Therefore, when the generation of Sutherland, Nelson, and Kay started to create "new media," they built it on top, so to speak, of what computers were already known to be capable of. Consequently they added new properties into physical media they were simulating right away. This can be very clearly seen in the case of Sketchpad. Understanding that one of the roles a computer can play is that of a problem solver, Sutherland built in a powerful new feature that never before existed in a graphical medium—satisfaction of constraints. To rephrase this example in more general terms, we can say that rather than moving from an imitation of older media to finding its own language, computational media was from the very beginning speaking a new language.

In other words, the pioneers of computational media did not have the goal of making the computer into a 'remediation machine" which would simply represent older media in new ways. Instead, knowing well the new capabilities provided by digital computers, they set out to create fundamentally new kinds of media for expression and communication. These new media would use as their raw "content" the older media which already served humans well for hundreds and thousands of years—written language, sound, line drawings and design plans, and continuous tone images (i.e. paintings and photographs). But this does not compromise the newness of new media. Computational media uses these traditional human media simply as building blocks to create previously unimaginable representational and information structures, creative and thinking tools, and communication options.

Although Sutherland, Engelbart, Nelson, Kay, and others developed computational media on top of already existing developments in computational theory, programming languages, and computer engineering, it would be incorrect to conceive the history of such influences as only going in one direction—from already existing and more general computing principles to particular techniques of computational media. The inventors of computational media had to question many, if not most, already established ideas about computing. They have defined many new fundamental concepts and techniques of how both software and hardware function, thus making important contributions to hardware and software engineering. A good example is Kay's development of Smalltalk, which for the first time systematically established a paradigm of object-oriented programming. Kay's rationale to develop this new programming language was to give a unified appearance to all applications and the interface of the PARC system and, even more importantly, to enable its users to quickly program their own media tools. (According to Kay, an object-oriented illustration program written in Smalltalk by a particularly talented 12-year-old girl was only a page long.[48]) Subsequently the object-oriented programming paradigm became very popular and object-oriented features have been added to most popular languages such as C.

Looking at the history of computer media and examining the thinking of its inventors makes it clear that we are dealing with the opposite of technological determinism. When Sutherland designed Sketchpad, Nelson conceived hypertext, Kay programmed a paint program, and so on, each new property of computer media had to be imagined, implemented, tested, and refined. In other words, these characteristics did not simply come as an inevitable result of a meeting between digital computers and modern media.

---

[48] Alan Kay, *Doing with Images Makes Symbols* (University Video Communications, 1987), videotaped lecture, http://archive.org/details/AlanKeyD1987/

Computational media had to be invented, step-by-step. And it was invented by people who were looking for inspiration in modern art, literature, cognitive and education psychology, and theory of media as much as technology. For example, Kay recalls that reading McLuhan's *Understanding Media* led him to a realization that a computer can be a medium rather than only a tool.[49] Accordingly, the opening section of Kay and Goldberg's article is called "Humans and Media," and it does read like media theory. But this is not a typical theory that only describes the word, as it currently exists. Similar to Marx's analysis of capitalism in his works, here the analysis is used to create a plan for action for building a new world—in this case, enabling people to create new media.

But the most important example of such non-deterministic development is the invention of the modern interactive graphical human-computer interface itself by Sutherland, Engelbart, Kay and others. None of the key theoretical concepts of modern computing as developed by Turing and Von Neumann called for an interactive interface. In the late 1940s and 1950s the MIT Lincoln Laboratory developed interactive graphical computers used in SAGE—the control centers created around the US to collect information from radar stations and coordinate a counter-attack. But the SAGE interface was designed for very particular tasks and it had no effect on the development of commercial computing. It did, however, lead to a new smaller machine: the TX-2, used by young students at MIT (including Sutherland) to explore what can be done with an "interactive computer"—i.e. a computer which had a visual display. Some students started to create interactive games including the famous Spacewar (1960). Sutherland was one of these students who were exploring the possibilities of visual interactive computing using the TX-2. He went to create Sketchpad (his Ph.D. thesis) which influenced other pioneers of cultural computing in the 1960s including Kay. But the theoretical road that led from SAGE to modern GUI through PARC was a very long one.

According to Kay, the key step for him and his group was to start thinking about computers as a medium for learning,

---

[49] Alan Kay, "User Interface: A Personal View," p. 192–3.

experimentation, and artistic expression which can be used not just by adults but also by "children of all ages."[50] Kay was strongly influenced by the theory of the cognitive psychologist Jerome Bruner. Bruner developed his theory by redefining the ideas of Jean Piaget who postulated that children go through a number of distinctive intellectual stages as they develop: a kinesthetic stage, a visual stage, and a symbolic stage. But while Piaget thought that each stage only exists for a particular period during a child's development only to be completely replaced by a new stage, Bruner suggested that separate mentalities that correspond to these stages continue to exist as the child grows. That is, the mentalities do not replace each other but are added. Bruner gave slightly different names to these different mentalities: enactive, iconic, and symbolic. While each mentality has developed at different stages of human evolution, they continue to co-exist in an adult.

Kay's interpretation of this theory was that a user interface should appeal to all these three mentalities. In contrast to a command-line interface, which is not accessible for children and forces the adult to use only symbolic mentality, the new interface should also make use of emotive and iconic mentalities. Kay also drew on a number of studies on creativity in math, science, music, art and other areas which suggested that initial creative work is done mostly in iconic mentality and also in enactive.[51] This provided additional motivation for the idea that if computers were to function as a dynamic medium for learning and creativity they should allow their users to think not only through symbols but also through actions and images.

Following Kay's interpretation of Bruner's work, the group at PARC mapped Bruner's theory of multiple mentalities into the interface technologies in the following way. *Mouse* activates enactive mentality (know where you are, manipulate). *Icons and windows* activate iconic mentality (recognize, compare, configure.) Finally, *Smalltalk programming language* allows for the use

---

[50] Alan Kay, "A Personal Computer for Children of All Ages," *Proceedings of the ACM National Conference*, Boston, 1972, http://www.mprove.de/diplom/gui/kay72.html

[51] Alan Kay, "User Interface: A Personal View," p. 195.

of symbolic mentality (tie together long chains of reasoning, abstract.)[52]

In actual use, a contemporary GUI involves constant interplay between different mentalities. You use a mouse to move around the screen as though it is a physical space and point at screen objects. All objects are represented by visual icons. You double-click on an icon to activate it or, if it is a folder icon, to examine its contents. This can be interpreted as an equivalent of picking up and examining a physical object in a real world. After a folder window opens, you may switch between different views, looking at the data as icons and alternatively as a list, then sort the list in different ways to examine file names, creation dates and other symbolic information (i.e. text). If you did not find the files you were looking for, you may then use a search function to search the whole computer—possibly defining multiple options and carefully choosing the search terms (symbolic mentality). As these examples demonstrate, the user is constantly switching between different mentalities using whatever works best at a given moment.

But in addition to the general interface principles, other key techniques that were developed by Kay's group can also be understood as enabling the use of different mentalities in combination with each other. For instance, the user interface developed at PARC was the first to run on a bit-mapped display—which meant not only giving users the ability to move the pointer and open multiple windows but also to write simulation programs in Smalltalk which could display their results visually right on the screen. By making a change in the code a user would be able to see the visual result of this change in the image produced by the program. Today this ability is fundamental to computer use in all areas of science (in particular, the use of interactive visualization and data analysis software). And of course, we should not forget about all the media editors created at PARC: a paint program, an illustration program, a music editor, etc. These media editors gave the users the ability to switch between different mentalities in a way not available in the physical media. For instance, the objects in the animation program could be drawn by hand or by writing code in Smalltalk. As Kay and Goldberg point out, "The control of the animation

[52] *Ibid.*, p. 197.

could be easily done from a Smalltalk simulation. For example, an animation of objects bouncing in a room is most easily accomplished by a few lines of Smalltalk code that express the class of bouncing objects in physical terms."[53]

In defining this new type of user interface, Kay and his collaborators simultaneously created a radically new type of media. If we are to agree with Bruner's theory of multiple mentalities and Kay's interpretation of this theory, we should conclude that the new computational media that he helped to invent can do something no previous media can—activate our multiple mentalities which all play a role in learning and creativity, allowing a user to employ whatever works best at any given moment and to rapidly switch between them as necessary. This may explain the success and popularity of the GUI, which, forty years after its invention, continues to dominate our interaction with computers. People prefer it not because it is "easy" or "seamless" or "intuitive." It is successful because it was designed to help them think, discover, and create new concepts using not just one type of mentality but all of them together. In short, while many HCI experts and designers continue to believe that the ideal human-computer interface should be invisible and get out of the way to let users do their work, looking at the theories of Kay and Goldberg that were behind GUI design gives a very different way of understanding an interface's identity. Kay and his colleagues at PARC have conceived GUI as a *medium* designed in its every detail to facilitate learning, discovery, and creativity.

Given the overall emphasis of information society on constant innovation, continuous learning, and creativity, it is only appropriate that as this society was coming into existence, a new medium was being invented specifically to facilitate these needs. In 1973 Daniel Bell published his highly influential *The Coming of Post-Industrial Society*; right around that time at PARC Kay, Goldberg, Chuck Thacker, Dan Ingalls, Larry Tesler, and other members of the Learning Research Group created the paradigm of modern computing. Or rather, they reinvented the computer—from a fast calculator that can only work on tasks articulated

[53] Kay and Goldberg, "Personal Dynamic Media," p. 399.

beforehand to an interactive support system for thinking and discovery. In short: from a tool to a metamedium.

Unfortunately, when GUI became the commercially successful paradigm following the success of Apple's Mac computers, introduced in 1984, the intellectual origins of GUI were forgotten. Instead, GUI was justified using a simplistic idea that since computers are unfamiliar to people, we should help them by making interface intuitive by making it mimic something users are already well familiar with—the physical world outside of a computer (which in reality was an office environment with folders, desks, printers, etc.) Surprisingly, even in recent years– when "born digital" generations were already using computer devices even before they ever set foot in an office—this idea was still used to explain GUI. For example, Apple's iPhone Human Interface guidelines (March 2010) advise developers: "When possible, model your application's objects and actions on objects and actions in the real world. This technique especially helps novice users quickly grasp how your application works. Folders are a classic software metaphor. People file things in folders in the real world, so they immediately understand the idea of putting data into folders on a computer."[54] The irony of this statement is that these Interface guidelines are also aimed at the developers of iPad—which clearly represents yet another step in migration from the world of physical print to all-digital environment. It is as though we are asked to remember and cherish the older media—and erase it at the same time.

The computer as a metamedium

As we have established, the development of computational media runs contrary to previous media history. But in a certain sense, the idea of a new media gradually discovering its own language actually does apply to the history of computational media after all. And just as with printed books and cinema, this process took a

[54] http://developer.apple.com/iphone/library/documentation/UserExperience/Conceptual/MobileHIG/PrinciplesAndCharacteristics/PrinciplesAndCharacteristics.html#//apple_ref/doc/uid/TP40006556-CH7-SW1 (April 5, 2010).

few decades. When the first computers were built in the middle of the 1940s, they could not be used as media for cultural representation, expression, and communication. Slowly, through the work of Sutherland, Engelbart, Nelson, Papert, and others in the 1960s, the ideas and techniques were developed that made computers into a cultural machine. One could create and edit text, make drawings, move around a virtual object, etc. And finally, when Kay and his colleagues at PARC systematized and refined these techniques and put them under the umbrella of a GUI (making computers accessible to multitudes) a digital computer finally was given its own language—in cultural terms. In short, only when a computer became a cultural medium—rather than merely a versatile machine—could it be so used.

Or rather, it became something that no other media had been before. For what had emerged was not yet another media, but as Kay and Goldberg insist in their article, something qualitatively different and historically unprecedented. To mark this difference, they introduce a new term—"metamedium."

This metamedium is unique in a number of different ways. One of them I have already discussed in detail—it can represent most other media while augmenting them with many new properties. Kay and Goldberg also name other properties that are equally crucial. The new metamedium is "active—it can respond to queries and experiments—so that the messages may involve the learner in a two-way conversation." For Kay who was strongly interested in children and learning, this property was particularly important since, as he puts it, it "has never been available before except through the medium of an individual teacher." [55] Further, the new metamedium can handle "virtually all of its owner's information-related needs." (I have already discussed the consequence of this property above.) It can also serve as "a programming and problem solving tool" and "an interactive memory for the storage and manipulation of data."[56] But the property that is the most important from the point of view of media history is that *the computer metamedium is simultaneously a set of different media and a system for generating new media tools and new types of media.* In other words, a computer

[55] Kay and Goldberg, "Personal Dynamic Media," p. 394.
[56] *Ibid.*, p. 393.

can be used to create *new tools for working with the media types it already provides as well as to develop new not-yet-invented media.*

In the opening to his book *Expressive Processing*, Noah Wardrip-Fruin perfectly articulates this "meta-generative" specificity of computers:

> A computer can simulate a typewriter—getting input from the keyboard and arranging pixels on the screen to shape the corresponding letters—but it can also go far beyond a typewriter, offering many fonts, automatic spelling correction, painless movement of manuscript sections (through simulations of "cut" and "paste"), programmable transformations (such as "find and replace"), and even collaborative authoring by large, dispersed groups (as with projects like Wikipedia). This is what modern computers (more lengthily called "stored-program electronic digital computers") are designed to make possible: the continual creation of new machines, opening new possibilities, through the definition of new sets of computational processes.[57]

Using the analogy with print literacy, Kay motivates this property in this way: "The ability to 'read' a medium means you can *access* materials and tools generated by others. The ability to write in a medium means you can *generate* materials and tools for others. You must have both to be literate."[58] Accordingly, Kay's key effort at PARC was the development of the Smalltalk programming language. All media editing applications and the GUI itself were written in Smalltalk. This made all the interfaces of all applications consistent, facilitating quick learning of new programs. Even more importantly, according to Kay's vision, Smalltalk would allow even novice users to write their own tools and define their own media. In other words, all media editing applications that would be provided with a computer, were to serve also as examples, inspiring users to modify them and to write their own applications.

---

[57] Noah Wardrip-Fruin, *Expressive Processing: Digital Fictions, Computer Games, and Software Studies* (The MIT Press, 2009).

[58] Alan Kay, "User Interface: A Personal View," in *The Art of Human-Computer Interface Design*, ed. Brenda Laurel (Reading, MA, Addison-Wesley, 1990), p. 193. The emphasis is in the original.

Accordingly, the large part of Kay and Goldberg's paper is devoted to description of software developed by the users of their system: "an animation system programmed by animators", "a drawing and painting system programmed by a child," "a hospital simulation programmed by a decision-theorist," "an audio animation system programmed by musicians", "a musical score capture system programmed by a musician", "electronic circuit design by a high school student." As can be seen from this list, (which corresponds to the sequence of examples in the article), Kay and Goldberg deliberately juxtaposed different types of users—professionals, high school students, and children—in order to show that everybody could develop new tools using the Smalltalk programming environment.

The sequence of examples also strategically juxtaposes media simulations with other kinds of simulations in order to emphasize that simulation of media is only a particular case of the computer's general ability to simulate all kinds of processes and systems. This juxtaposition of examples gives us an interesting way to think about computational media. Just as a scientist may use simulation to test different conditions and play different what/if scenarios, a designer, a writer, a musician, a filmmaker, or an architect working with computer media can quickly "test" different creative directions in which the project can be developed as well as see how modifications of various "parameters" affect the project. The latter is particularly easy today since the interfaces of most media editing software not only explicitly present these parameters but also simultaneously give the user the controls for their modification. For instance, when the Formatting Palette in Microsoft Word shows the font used by the currently selected text, it is displayed in a column next to all other fonts available. Trying different fonts is as easy as scrolling down and selecting the name of a new font.

Giving users the ability to write their own programs was a crucial part of Kay's vision for the new "metamedium" he was inventing at PARC. According to Noah Wardrip-Fruin, Engelbart's research program was focused on a similar goal: "Engelbart envisioned users creating tools, sharing tools, and altering the tools of others."[59]

[59] Noah Wardrip-Fruin, introduction to Douglas Engelbart and William English, "A Research Center for Augmenting Human Intellect" (1968), *New Media Reader*, p. 232.

Unfortunately, when in 1984 Apple shipped Macintosh, which was to become the first commercially successful personal computer modeled after the PARC system, it did not have an easy-to-use programming environment. HyperCard, written for Macintosh in 1987 by Bill Atkinson (who was one of PARC's alumni), gave users the ability to quickly create certain kinds of applications—but it did not have the versatility and breadth envisioned by Kay. Only more recently, as the general computer literacy has widened and many new high-level programming languages have become available—Perl, PHP, Python, JavaScript, etc.—have more people started to create their own tools by writing software. A good example of a contemporary programming environment, very popular among artists and designers and which, in my view, is close to Kay's vision, is Processing.[60] Built on top of the Java programming language, Processing features a simplified programming style and an extensive library of graphical and media functions. It can be used to develop complex programs and also to quickly test ideas. Appropriately, the official name for Processing projects is sketches.[61] In the words of Processing inventors and main developers Ben Fry and Casey Reas, the language's focus is "on the 'process' of creation rather than end results."[62] Another popular programming environment that similarly enables quick development of media projects is Max/MSP and its successor PD—both developed by Miller Puckette.

At the end of the 1977 article that served as the basis for our discussion in this chapter, Kay and Goldberg summarize their arguments in the phrase—which in my view is the best formulation we have had so far—of what computational media is artistically and culturally. They call the computer *"a metamedium"* whose content is *"a wide range of already-existing and not-yet-invented media."* In another article published in 1984 Kay unfolds this definition. As a way of concluding this chapter, I would like to quote this longer definition which is as accurate and inspiring today as it was when Kay wrote it:

> It [a computer] is a medium that can dynamically simulate the details of any other medium, including media that cannot exist

[60] www.processing.org

[61] http://www.processing.org/reference/environment/

[62] http://wiki.processing.org/w/FAQ

physically. It is not a tool, though it can act like many tools. It is the first *metamedium*, and as such it has degrees of freedom for representation and expression never before encountered and as yet barely investigated.[63]

[63] Alan Kay, "Computer Software," *Scientific American* (September 1984), p. 52. Quoted in Jean-Louis Gassée, "The Evolution of Thinking Tools," in *The Art of Human-Computer Interface Design*, p. 225.

CHAPTER TWO

Understanding metamedia

"It [the electronic book] need not be treated as a simulated paper book since this is a new medium with new properties."
Kay and Goldberg, "Personal Dynamic Media," 1977

Today *Popular Science*, published by Bonnier and the largest science+tech magazine in the world, is launching Popular Science+ — the first magazine on the Mag+ platform, and you can get it on the iPad tomorrow...What amazes me is that you don't feel like you're using a website, or even that you're using an e-reader on a new tablet device — which, technically, is what it is. *It feels like you're reading a magazine.*" (emphasis is in the original.)
"Popular Science+," posted on April 2, 2010.
http://berglondon.com/blog/2010/04/02/popularscienceplus/

The building blocks

I started putting this book together in 2007. Today is April 3, 2010, and I am editing this chapter. Today is also an important day in the history of media computing (which started exactly forty years ago with Ivan Sutherland's Sketchpad)—Apple's iPad tablet computer first went on sale in the US on this date. During the years I was writing and editing the book, many important developments made Alan Kay's vision of a computer as the "first metamedium" more real—and at the same time more distant.

The dramatic cuts in the prices of laptops and the rise of cheap notebooks (and, in the years that followed, tablet computers), together with the continuing increase in the capacity and decrease in price of consumer electronics devices (digital cameras, video cameras, media players, monitors, storage, etc.) brought media computing to even more people. With the price of a notebook ten or twenty times less than the price of a large digital TV set, the 1990s' argument about the "digital divide" became less relevant. It became cheaper to create your own media than to consume professional TV programs via the industry's preferred mode of distribution. More students, designers, and artists learned Processing and other specialized programming and scripting languages specifically designed for their needs—which made software-driven art and media design more common. Perhaps most importantly, most mobile phones became "smart phones" supporting Internet connectivity, web browsing, email, photo and video capture, and a range of other media creation capabilities—as well as the new platforms for software development. For example, Apple's iPhone went on sale on June 29, 2007; on July 10 when the App Store opened, it already had 500 third-party applications. According to Apple's statistics, on March 20, 2010 the store had over 150,000 different applications and the total number of application downloads had reached 3 billion. In February 2012, the numbers of iOS apps reached 500,000 (not counting many more that Apple rejected), and the total number of downloads was already 25 billion.[1]

At the same time, some of the same developments strengthened a different vision of media computing—a computer as a device for buying and consuming professional media, organizing personal media assets and using GUI applications for media creation and editing—but not imagining and creating "*not-yet-invented media.*" Apple's first Mac computer, released in 1984, did not support writing new programs to take advantage of its media capacities. The adoption of the GUI interface for all PC applications by the software industry made computers much easier to use but in the same time took away any reason to learn programming. Around 2000, Apple's new paradigm of a computer as a "media hub" (or a "media center")—a platform for managing all personally created

---

[1] http://www.apple.com/itunes/25-billion-app-countdown/ (March 5, 2012).

media—further erased the "computer" part of a PC. During the following decade, the gradual emergence of web-based distribution channels for commercial media, such as Apple iTunes Music Store (2003), internet television (in the US the first successful service was Hulu, publically launched on March 12, 2008), the e-book market (Random House and HarperCollins started selling their titles in digital form in 2002) and finally the Apple iBookstore (April 3, 2010), together with specialized media readers and players such as Amazon's Kindle (November 2007) have added a new crucial part to this paradigm. (In the early 2010s we also got the Android app market, the Amazon app store, etc.) A computer became even more of a "universal media machine" than before—with the focus on consuming media created by others.

Thus, if in 1984 Apple's first Apple computer was critiqued for its GUI applications and lack of programming tools for the users, in 2010 Apple's iPad was critiqued for not including enough GUI tools for heavy duty media creation and editing—another step back from Kay's Dynabook vision. The following quote from an iPad review by Walter S. Mossberg from the *Wall Street Journal* was typical of journalists' reactions to the new device: "if you're mainly a Web surfer, note-taker, social-networker and emailer, and a consumer of photos, videos, books, periodicals and music—this could be for you."[2] The *New York Times*' NYT's David Pogue echoed this: "The iPad is not a laptop. It's not nearly as good for creating stuff. On the other hand, it's infinitely more convenient for consuming it—books, music, video, photos, Web, e-mail and so on."[3]

Regardless of how much contemporary "universal media machines" fulfill or betray Alan Kay's original vision, they are only possible because of it. Kay and others working at Xerox PARC built the first such machine by creating a number of media authoring and editing applications with a unified interface, as well as the technology to enable the machine's users to extend its capacities. Starting with the concept Kay and Goldberg proposed in 1977, to sum up this work at PARC (the computer as *"a metamedium"* whose content is *"a wide range of already-existing and not-yet-invented media"*) in this chapter I will discuss how this

[2] http://ptech.allthingsd.com/20100331/apple-ipad-review/ (April 3, 2010).
[3] http://gizmodo.com/5506824/first-ipad-reviews-are-in (April 3, 2010).

concept redefines what media is. In other words, I will go deeper into the key question of this book: what exactly is media after software?

Approached from the point of view of media history, the computer metamedium contains two different types of media. The first type is *simulations of prior physical media extended with new properties*, such as "electronic paper." The second type is a number of *new computational media that have no physical precedents*. Here are the examples of these "new media proper," listed with names of the people and/or places usually credited as their inventors: hypertext and hypermedia (Ted Nelson); interactive navigable 3D spaces (Ivan Sutherland), interactive multimedia (Architecture Machine Group's "Aspen Movie Map").

This taxonomy is consistent with the definition of the computer metamedium given in the end of Kay and Goldberg's article. But let us now ask a new question: what are the building blocks of these simulations of previously existing media and newly invented media? Actually we have already encountered these blocks in the preceding discussion but until now I have not explicitly pointed them out.

The building blocks used to make up the computer metamedium are *different types of media data* and the *techniques for generating, modifying, and viewing this data*. Currently, the most widely used data types are text, vector images and image sequences (vector animation), continuous tone images and sequences of such images (i.e., photographs and digital video), 3D models, geo-spatial data, and audio. I am sure that some readers would prefer a somewhat different list and I will not argue with them. What is important at this point for our discussion is to establish that we have multiple kinds of data rather than just one kind.

This points leads us to the next one: the techniques for data manipulation themselves can be divided into two types depending on which data types they can work on:

(A) The first type is *media creation, manipulation, and access techniques that are specific to particular types of data*. In other words, these techniques can be used only on a particular data type (or a particular kind of "media content"). I am going to refer to these techniques as *media-specific* (the word "media" in this case really stands for "data type"). For example, the technique of geometrical constraint satisfaction invented by Sutherland can work on graphical data defined by points and lines. However, it would be meaningless to apply this technique to text. Another example: today image editing programs usually include various filters such as "blur" and "sharpen" which can operate on continuous tone images. But normally we would not be able to blur or sharpen a 3D model. Similarly, it would be as meaningless to try to "extrude" a text or "interpolate" it as to define a number of columns for an image or a sound composition.

Some of these data manipulation techniques appear to have no historical precedents in physical media—the technique of geometric constraint satisfaction is a case in point. Another example of such new technique is evolutionary algorithms commonly used to generate still images, animations, and 3D forms. Other media-specific techniques do refer to prior physical tools or machines—for instance, brushes in image editing applications, a zoom command in graphics software, or a trim command in video editing software. In other words, the same division between simulations and "properly new" media also applies to the individual techniques that make up the "computer metamedium."

(B) The second type is *new software techniques that can work with digital data in general (i.e. they are not media-specific)*. The examples are "view control," hyperlinking, sort, search, network protocols such as HTTP, and various data analysis techniques from the fields of Artificial Intelligence, Machine Leaning, Knowledge Discovery, and other sub-fields of computer science. (In fact, large parts of computer science, information science and computer engineering science are about these techniques—since they focus on designing algorithms for processing information in general.) These techniques are general ways of manipulating data regardless of what this data encodes (i.e. pixel values, text characters, sounds, etc.). I will be referring to these techniques as *media-independent*. For instance, as we saw, Engelbart's "view control"—the idea that the same

> information can be displayed in many different ways—is now implemented in most media editors and therefore works with images, 3D models, video files, animation projects, graphic designs, and sound compositions. "View control" has also become part of the modern OS (operating systems such as Mac OS X, Microsoft Windows, or Google Chrome OS). We use view control daily when we change the files "view" between "icons," "list," and "columns" (these are names used in Mac OS X; other OS may use different names to refer to the same views). General media-independent techniques also include interface commands such as cut, copy, and paste. For instance, you can select a file name in a directory, a group of pixels in an image, or a set of polygons in a 3D model, and then cut, copy, and paste these selected objects.

OK: we now have two different ways of "dividing up" the computer metamedium. If we want to continue using the concept of a "medium," we will say that a computer simulates prior mediums and allows for the definition of new ones. Alternatively, we can think of the computer metamedium as a collection of data types, media-specific techniques that can only operate on particular types, and media-independent techniques that can work on any data. Each of the mediums contained in the computer metamedium is made from some of these building blocks. For example, the elements of the "navigable 3D space" medium are 3D models plus techniques for representing them in perspective, texture mapping, simulating effects of various types of lights on their surface, casting shadows, and so on. In another example, the elements of "digital photography" are continuous tone images captured by lens-based sensors plus a variety of techniques for manipulating these images: changing contrast and saturation, scaling, compositing, etc.

A note about the distinction between media-specific and media-independent techniques: it works in theory. In practice, however, it is often hard to say in what category a particular technique or a medium should be placed. For instance, is Sketchpad's ability to work at any of 2000 levels of magnification an extension of techniques which existed previously (moving one's body closer to the drawing board, using a magnifying lens) or is it something really new? Or what about 3D navigable space, which I have used as an example of a new medium only made possible by computers (tracing it to Sutherland's first Virtual Reality system of 1966)? Is it new—or is it an extension of a physical medium of architecture, which allows a human being to walk around the built structures?

The boundaries between "simulated media" and "new media," or between "media-specific" and "media-independent" techniques should not be thought of as solid walls. Rather than thinking of them as rigidly defined categories, let us imagine them as coordinates of the map of the computer metamedium. Like any first sketch, no matter how imprecise, this map is useful because now we have something to modify as we go forward.

Media-independent *vs.* media-specific techniques

Having drawn our first map of the computer metamedium, let us now examine it to see if it can reveal something which we did not notice so far. We see a number of mediums, old and new—and this certainly fits with our common understanding of media history. (For example, take a look at the table of contents of McLuhan's *Understanding Media* and you will find two dozen chapters each devoted to a particular medium—which for McLuhan range from writing and roads to cars and TV.) We also see various media-specific techniques and this again is something we are familiar with: think of editing techniques in cinema, defining a contour in painting, creating rhyme in poetry, or shaping a narrative out of chronological story events in literature. But one area of the map does looks new and different in relation to previous cultural history. This is the area that contains "media-independent techniques." What are these techniques, and how can they work across media, i.e. on different types of data? ("Design across media" was a phrase used by Adobe in marketing an early version of its Creative Suite of media authoring applications.)

I am going to argue that "media independence" does not just happen by itself. For a technique to work with various data types, programmers have to implement a different method for each data type. Thus, *media-independent techniques are general concepts translated into algorithms, which can operate on particular data types*. Let us look at some examples.

Consider the omnipresent *cut* and *paste*. The algorithm to select a word in a text document is different from the algorithm to select a curve in a vector drawing, or the algorithm to select a part of a continuous tone (i.e. raster) image. In other words, "cut and paste" is a general concept that is implemented differently in different media software depending on which data type this software is designed to handle. (In Larry Tesler's original implementation of the universal commands concept done at PARC in 1974–5, it only worked for text editing.) Although cut, copy, paste, and a number of similar "universal commands" are available in all contemporary GUI applications for desktop computers (but not necessarily in mobile phone apps), what they actually do and how they do it is different from application to application.

Search operates in the same way. The algorithm to search for a particular phrase in a text document is different than the algorithm that searches for a particular face in a photo or a video clip. (I am talking here about "content-based search," i.e. the type of search which looks for information inside actual images, as opposed to only searching image titles and other metadata the way image search engines such as Google Image Search were doing it in the 2000s.) However, despite these differences the general concept of search is the same: locating any elements of a single media object—or any media objects in a larger set—to match particular user-defined criteria. Thus we can ask the web browser to locate all instances of a particular word in a current web page; we can ask a web search engine to locate all web pages which contain a set of keywords; and we can ask a content-based image search engine to find all images that are similar in composition to an image we provided.

Because of the popularity of the search paradigm on the web, we now assume that in principle we can—or will be able to in the future—search any media. In reality it is much easier to search data that has a modular organization—such as text or 3D models—than media that does not have it, such as continuous-tone images, video, or audio. But for the users these differences are not important—as far as they are concerned, all types of media content acquire a new common property that can be called *searchability*.

Similarly, in the mid-2000s photo and video media started

to acquire another property of *findability*. (I am borrowing this term from 2005 book by Peter Morville, *Ambient Findability: What We Find Changes Who We Become*[4]). The appearance of consumer GPS-enabled media capture devices and the addition of geo-tagging, geo-search, and mapping services to media sharing sites such as Flickr (added in 2006) and media management applications such as iPhoto (added in 2009) gradually made media "location aware."

Another example of a general concept that, through the efforts by many people, was gradually made to work with different media types—and thus became a "media-independent technique"—is *information visualization* (often abbreviated as *infovis*.) The name "infovis" already suggests that it is not a media-specific technique—rather it is a very general method that potentially can be applied to any data. The name implies that we can potentially take anything—numbers, text, network, sound, video, etc.—and map it into image to reveal patterns and relationships in the data. (A parallel to information visualization is *data sonification*, which renders data as sound).

However, it took decades to invent techniques to turn this potential into reality. In the 1980s the emerging field of scientific visualization focused on 3D visualization of numerical data. In the second part of the 1990s the growing graphics capabilities of PCs made possible for larger numbers of people to experiment with visualization—which led to the development of techniques to visualize media. The first successful visualizations of large bodies of text appeared around 1998 (*Rethinking the Book* by David Small, 1998; *Valence* by Ben Fry, 1999; *TextArc* by W. Bradford Paley, 2002[5]); visualizations of musical structures in 2001 (*The Shape of Song* by Martin Wattenberg); and visualization of a feature film in 2000 *(The Top Grossing Film of All Time, 1 x 1* by Jason Salovan).

Information visualization is a particularly interesting example of a new "media-independent technique" because of the variety of the algorithms and strategies for representing data visually.

[4] Peter Morville. *Ambient Findability: What We Find Changes Who We Become.* O'Reilly Media, Inc., 2005.

[5] W. Bradford Paley, TextArc, 2002, http://www.textarc.org/; Ben Fry, *Valence*, 1999, http://benfry.com/valence/; David Small, *Rethinking the Book*, PhD thesis, 1999, http://acg.media.mit.edu/projects/thesis/DSThesis.pdf

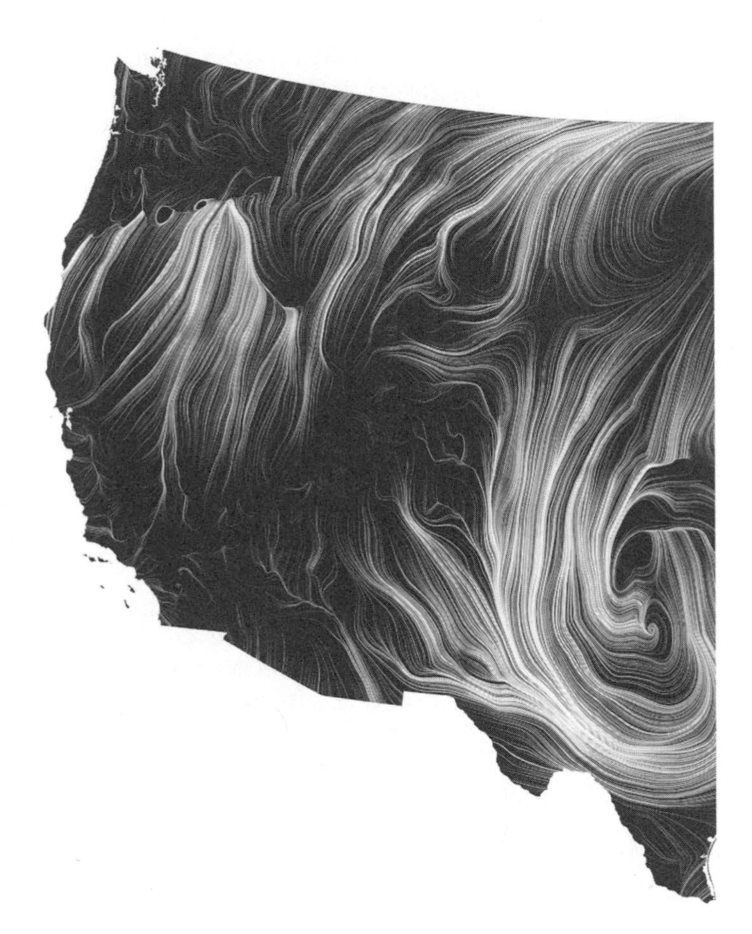

Wind Map, *a real-time dynamic visualization of the wind currents over the USA. Fernanda Viégas and Martin Wattenberg, 2012.*

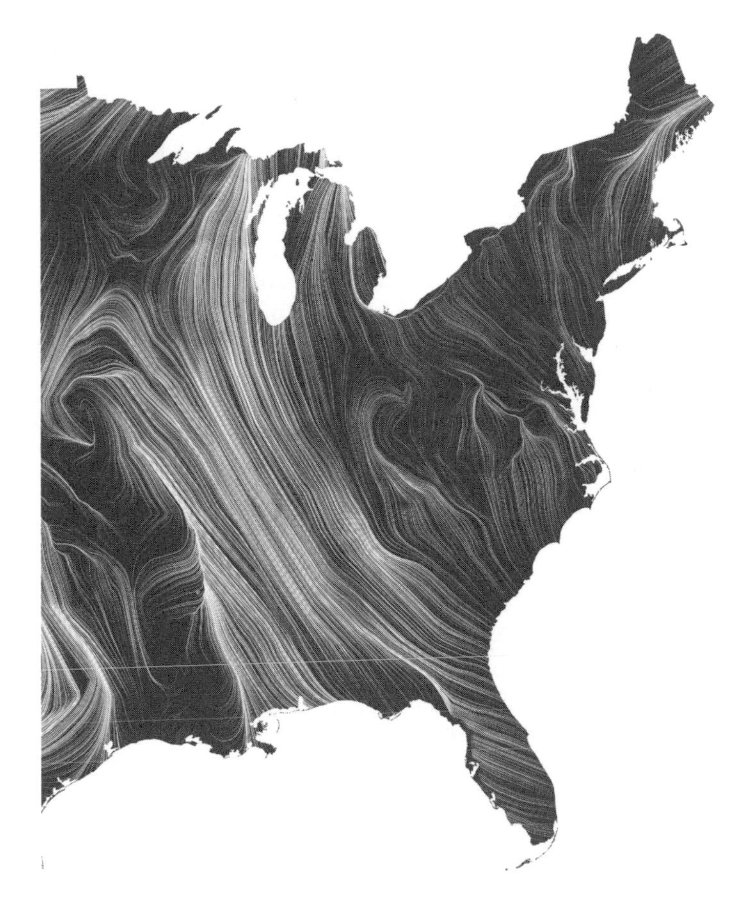

For example, Martin Wattenberg whose work, in his own words, "focuses on visual explorations of culturally significant data,"[6] created visualizations of a history of net art, the music compositions of Bach, Philip Glass and other composers, the thought process of a computer chess-playing program, and the history of Wikipedia pages, among other projects. In each case he had to decide which dimensions of the data to choose and how to translate them in a visual form. But despite the differences, we recognize all these projects as information visualizations. They are all realizations of the same general concept—selecting some dimensions of the data and representing them visually through the relations of graphic elements.[7] They also all rely on the same fundamental capacities of software to manipulate numerical data and to map it from one form to another. Finally, they all can be also understood as the application of the new computer media of computer graphics—generation of images from numerical data. (Think of an affinity between a 3D computer model based on a 3D scan of a face, and a vector visualization of, for instance, the face's position over time, based on the data extracted from a video.)

As the result of infovis work by Wattenberg and other people over the course of last two decades many types of data acquired a new common property—their structure can be visualized. This new property of media is distributed across various applications, software libraries, art and design projects, research papers, and prototypes. Today some visualization tools are included in media editing software—for instance, media editors such as Photoshop can display a histogram of an image, Final Cut and other professional video editing software can visualize the color content of a video clip, and many media players including iTunes offer a music visualization feature. Google Trends visualizes search patterns; YouTube and Flickr visualize viewing stats for video and photos. Going through the thousands of infovis projects collected on infosthetics.com, visualcomplexity.com, and other blogs about visualizations, we find a variety of experiments in visualization of media such as songs, poems and novels and every possible kind

[6] http://www.bewitched.com/about.html (July 23, 2006).

[7] For a detailed discussion of infovis, most general principles and new developments, see my article "What is Visualization?" (2010), *Visual Studies* 26, no. 1 (March 2011).

of data—from the artist's son's, daughter's and cat's movements in their living room over a period of an hour (*1hr in front of the TV* by umblebee, 2008) to a citation network in science journals (*Eigenfactor.org* by Moritz Stefaner, 2009).[8] We can also find such projects in art exhibitions such as MOMA's 2008 *Design and Elastic Mind,*[9] SIGGRAPH 2009 *Info-Aesthetics,*[10] and MOMA's 2011 *Talk to Me*.

Visualization, searchability, findability—these and many other new "media-independent techniques" (i.e. concepts implemented to work across many data types) clearly stand out in the map of the computer metamedium we have drawn because they go against our habitual understanding of media as plural (i.e. as consisting of a number of separate mediums). If we can use the same techniques across different media types, what happens to these distinctions between mediums?

The idea that all artworks fall into a number of distinct mediums each with its own distinct techniques and representational devices was central to modern art and aesthetics. In his 1766 *Laokoon oder Über die Grenzen der Malerei und Poesie* (*Laocoon: An Essay on the Limits of Painting and Poetry)* German philosopher Gotthold Ephraim Lessing argued for the radical difference between poetry and painting since one is "extended" in time and the other is in space. The idea reached its extreme in the first two decades of the twentieth century when modernist artists focused their energy on discovering a unique language of each artistic medium. The following statement made in 1924 by Jean Epstein, a French avant-garde filmmaker and theoretician, is typical of modernist rhetoric of purity; countless statements like it appeared on the pages of avant-garde publications of the time:

> For every art builds its forbidden city, its own exclusive domain, autonomous, specific and hostile to anything that does not belong. Astonishing to relate, literature must first and foremost be literary; the theater, theatrical; painting, pictorial; and the cinema, cinematic. Painting today is freeing itself from many of

---

[8] http://well-formed.eigenfactor.org/; http://www.flickr.com/photos/the_bumblebee/2229041742/in/pool-datavisualization

[9] http://www.moma.org/interactives/exhibitions/2008/elasticmind/

[10] http://www.siggraph.org/s2009/galleries_experiences/information_aesthetics/

> its representational and narrative concerns... The cinema must seek to become, gradually and in the end uniquely, cinematic; to employ, in other words, only photogenic elements.[11]

In relation to painting, the doctrine of media purity reaches its extreme expression in the famous argument of Clement Greenberg that "Because flatness was the only condition painting shared with no other art, Modernist painting oriented itself to flatness as it did to nothing else."[12] (Note that Greenberg did not advocate this position as a justification for the abstract art of his contemporaries; he only offered this as a historical analysis of earlier modernism.)

Greenberg wrote: "It was the stressing of the ineluctable flatness of the surface that remained, however, more fundamental than anything else to the processes by which pictorial art criticized and defined itself under Modernism. For flatness alone was unique and exclusive to pictorial art."

Only after the 1960s when installation—a new art form based on the idea of mixing different media and materials—gradually became popular and accepted in the art world, did the obsession with media-specificity lose its importance.

However, even during its dominance the principle of media-specificity was always counterbalanced by its opposite. Throughout the modern period we also find "local"—i.e. specific to particular historical moments and artistic schools—attempts to formulate aesthetic principles that can relate different mediums to each other. Consider for instance the considerable efforts spent by many modernist artists to establish parallels between musical and visual compositions. This work was often associated with the ideas of synesthesia and Gesamtkunstwerk; it included theories, practical compositions, and technologies such as color organs constructed by Scriabin, the Whitneys (who went on to create the first computer animations) and many other artists and musicians.

While they did not explicitly theorize cross-media aesthetics to the same degree, modernist artistic paradigms—classicism,

---

[11] Jean Epstein, "On Certain Characteristics of *Photogénie*," in *French Film Theory and Criticism*, vol. 1: 1907–29, ed. Richard Abel (Princeton: University of Princeton Press, 1988).

[12] Clement Greenberg, "Modern Painting," *Forum Lectures* (Washington, DC: Voice of America: 1960), http://www.sharecom.ca/greenberg/modernism.html

romanticism, naturalism, impressionism, socialist realism, suprematism, cubism, surrealism, and so on—can be also understood as the systems which gave "common properties" to works in various media. Thus, the novels of Émile Zola and paintings by Manet were aligned in their uncompromising, "naturalist" depiction of ordinary people; Constructivist paintings, graphics, industrial design, theatre design, architecture and fashion shared the aesthetics of "expressed structure" (visually emphasizing composition structure by exaggerating it); and De Stijl aesthetics of non-intersecting rectangular forms in primary colors were applied to painting, furniture, architecture, and typography.

What is the difference between such earlier artistic work on establishing media correspondences and the software techniques that work across different media? Clearly, the artistic systems and media authoring, editing and interaction techniques available in media software operate on different levels. The former are responsible for the content and style of the works to be generated—i.e. what is going to be created in the first place. The latter are used not only to create but also to interact with what already was generated previously—for instance, blogs, photos, or videos on the web created by others.

Put differently, the efforts by modern artists to create parallels between mediums were prescriptive and speculative.[13] "Common media properties" would only apply to selected bodies of artistic work created by particular artists or groups. In contrast, software imposes common media "properties" on any media it is applied to. Thus, software also shapes our understanding of what media is in general. For example, web applications and services include methods for navigating, reading, listening or viewing media objects, attaching additional information to them (comments, tags, geo-tagging) or finding them in a larger set (i.e. search engines and search commands). This applies to all videos, images, text pages, text documents, maps, etc. In other words, we can say that media

---

[13] By speculative here I mean that in many cases the proposed aesthetic systems were not fully realized in practice. For example, no purely suprematist architecture designed by movement leader Kasimir Malevich was ever built; the same goes for the futurist architecture of Antonio Sant'Elia as presented in his drawings for *La Città Nuova*, 1912–14.

software "interprets" any media it touches and its "interpretations" always include certain statements.

Of course, "media-independent" aesthetic systems proposed by modernists were not only generative (the creation of new works) but also interpretive. That is, modernist artists and theorists often tried to change audiences' understanding of past and contemporary art, usually in a critical and negative way (each new movement wanted to discredit its predecessors and competitors). However, since their programs were theories rather than software, they had no direct material effect on users' interaction with the artistic works, including those created in the past. In contrast, software techniques affect our understanding of media through the operations they make available to us for creating, editing, interacting with, and sharing media artifacts.

Additionally, modern artistic and aesthetic paradigms in practice would only be realized in two, three, or maybe four mediums—but not all. Naturalism can be found in literature and visual arts but not in architecture, and Constructivism did not spread to music or literature. However cut, copy, paste and find commands are found in all media applications; any media object can be geo-tagged; the view control principle is implemented to work with all media types. To cite more examples, recall my discussion of how all media acquire new properties such as searchability and findability. Any text can be searched regardless of whether it is something you wrote or a classical novel downloaded from Project Gutenberg; similarly, part of an image can be cut and pasted into another image regardless of what these images are. In short: media software affects all media content equally, regardless of its aesthetics, semantics, authorship, and historical origin.

To summarize this discussion: in contrast to modern artistic programs to create different media that share the same principles, software media-independent techniques are ubiquitous and "universalist." For instance, cut and paste are built into all media editing software—from specialized professional applications to consumer software included on every new media device sold. Further, these techniques can be applied to any media work regardless of its aesthetics and authorship—i.e. whether it was made by the person who is currently applying these operations or by somebody else. In fact, the technical ability to sample media work by others has become the basis of the key aesthetics of our time—remixing.

Of course, not all media applications and devices make all these techniques equally available—usually for commercial and/or copyright reasons. For example, at present the Google Books reader does not allow you to select and copy the text from book pages. Thus, while our analysis applies to conceptual and technical principles of software and their cultural implications, we need to keep in mind that in practice these principles are overwritten by commercial structures. However, even the most restrictive software still includes some basic media operations. By being present in all software designed to work with different media types, these operations establish a shared understanding of media today—experienced by the users as "common media principles" of all content.

To conclude this discussion of software techniques that work across different media types, recall my earlier statement that computerization of media does not collapse the difference between mediums—but it does bring them closer together in various ways. (I have already provided a number of examples of this in the beginning of my analysis of Kay and Goldberg's *Personal Dynamic Media* article). Now I can name one of the key developments responsible for this "media attraction"—common software techniques which can operate across different media types. If we recall again Kay and Goldberg's formulation that the computer metamedium includes a variety of already-existing and new media, this statement can be paraphrased as follows: *within the computer metamedium, all previously existing and newly invented mediums share some common properties—i.e. they rely on a set of common software techniques for data management, authoring, and communication.*

It is hard to over-estimate the historical importance of the development of these cross-media techniques. Humans have always used some general strategies to organize their cultural representations, experiences and actions—for example, narrative, symmetry, rhythm, the repetitive structures (think of ornament), use of complementary colors, and a few others. Clearly these strategies were very important for human perception, cognition, and memory—and that is why we find them in every culture and every medium, from poetry to architecture, music and poetry. However, these strategies were normally not embedded into any technological tools or materials—they were in the minds and bodies of artisans who were communicating them from generation to generation. Modern media for representation and communication bring us to a new stage. They often do embody certain techniques that apply to any media which can be generated or captured with them (for instance, one-point linear perspective imposed by lens-based capture technologies such as photography, film and analog and digital video). However these techniques would apply only to particular media types. Against these historical developments, the innovation of media software clearly stands. They bring a new set of techniques which are implemented to work across all media. *Searchability*, *findability*, *linkability*, multimedia messaging and sharing, editing, view control, zoom and other "media-independent" techniques are viruses that infect everything software touches—and therefore in their importance they can be compared to the basic organizing principles for media and artifacts which were used for thousands of years.

Inside Photoshop

Contemporary media is experienced, created, edited, remixed, organized, and shared with software. This software includes stand-alone professional media design and management applications such as Photoshop, Illustrator, Dreamweaver, Final Cut, After Effects, Aperture, and Maya; consumer-level apps such as iPhoto, iMovie, or Picasa; tools for sharing, commenting, and editing provided by social media sites such as Facebook, YouTube, Vimeo, and Photobucket, and the numerous media viewing, editing, and sharing apps available on mobile platforms. To understand media today we need to *understand media software*—its genealogy (where it comes from), its anatomy (interfaces and operations), and its practical and theoretical effects. How does media authoring software shape the media being created, making some design choices seem natural and easy to execute, while hiding other design possibilities? How does media viewing/managing/remixing software affect our experience of media and the actions we perform on it? How does software change what "media" is conceptually?

This section continues investigating these general questions that drive this book via the analysis of a software application that became synonymous with "digital media"—Adobe Photoshop.

Like other professional programs for media authoring and editing, Photoshop's menus contain many dozens of separate commands. If we consider that almost all the commands contain multiple options that allow each command to do a number of different things, the complete number runs into the thousands.

This multiplicity of operations offered in contemporary application software creates a challenge for Software Studies. If we are to understand how software applications participate in shaping our worlds and our imaginations (what people imagine they can do with software), we need some way of sorting all these operations into fewer categories so we can start building a theory of application software. This cannot be achieved by simply following the top menu categories offered by applications. (For example, Photoshop CS4's top menu includes File, Edit, Layer, Select, Filter, 3D, View, Window, and Help.) Since most applications include their own unique categories, our combined list will be too large. So we need to use a more general system.

The provisional map of the computer metamedium that we developed in previous sections provides one such possible system. In this section we will test the usefulness of this map by analyzing a subset of Photoshop's commands which, in a certain sense, stand-in for this application in our cultural imagination: Filters. We will also discuss another key feature of Photoshop—Layers.

Our map organizes software techniques for working with media, using two schemes. The first scheme divides these techniques into two types depending on which data types they can work on: 1) media creation, manipulation, and access techniques that are specific to particular types of data; 2) new software techniques that can work with digital data in general (i.e. they are not specific to particular types of data). My second scheme also divides software techniques for working with media data into two types, but it does this in a different way. What matters here are the relations between software techniques and pre-digital media technologies. In this taxonomy, some techniques are simulations of pre-digital media techniques augmented with new properties and functions; other techniques do not have any obvious equivalents in previous physical or electronic media.

While for a media historian the second scheme is quite meaningful, what about users who are "digital natives"? These software users may never have directly used any other media besides tablets or

0.63

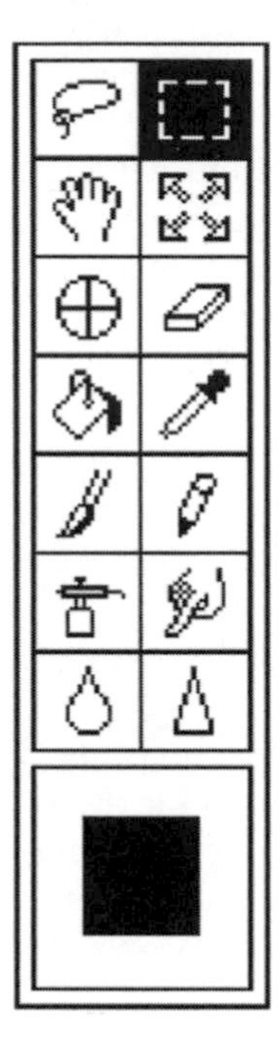

1.07

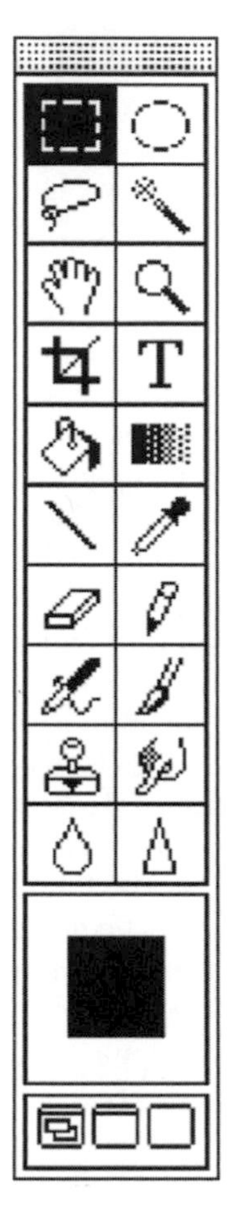

2.5 LE

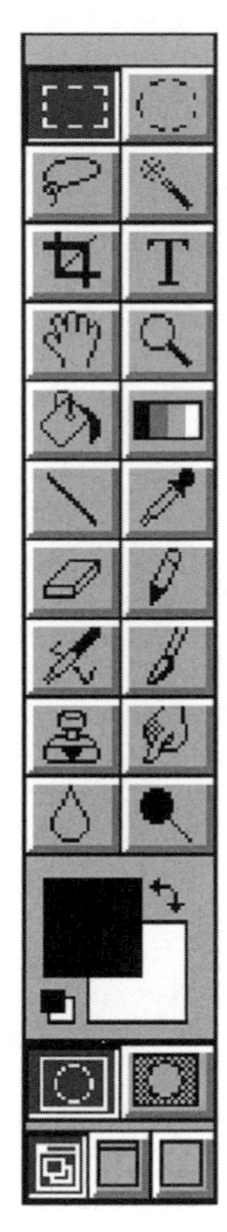

Photoshop Toolbox from version 0.63 (1988) to 7.0 (2002).

5.5

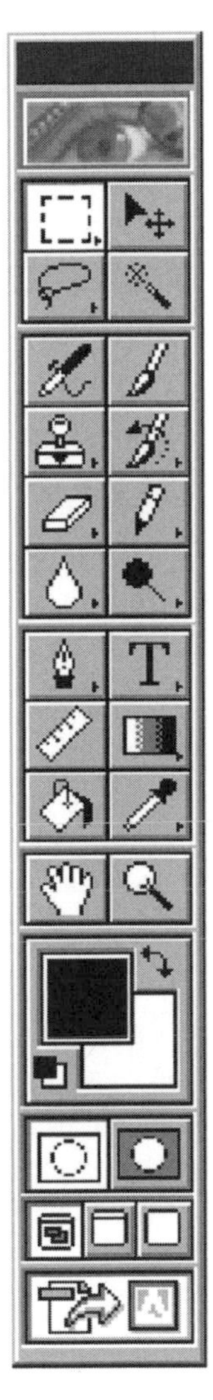

6.0

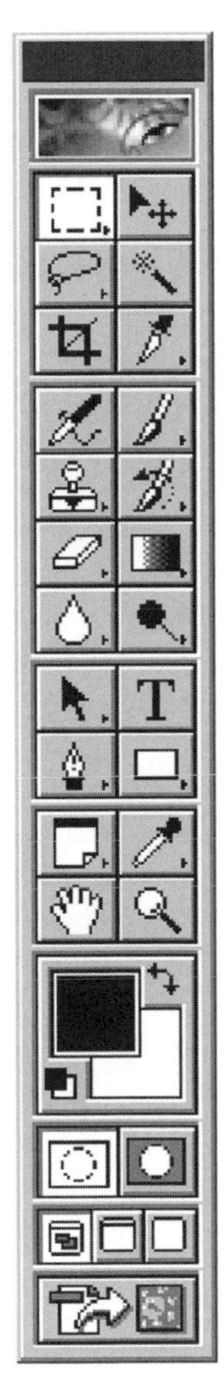

7.0

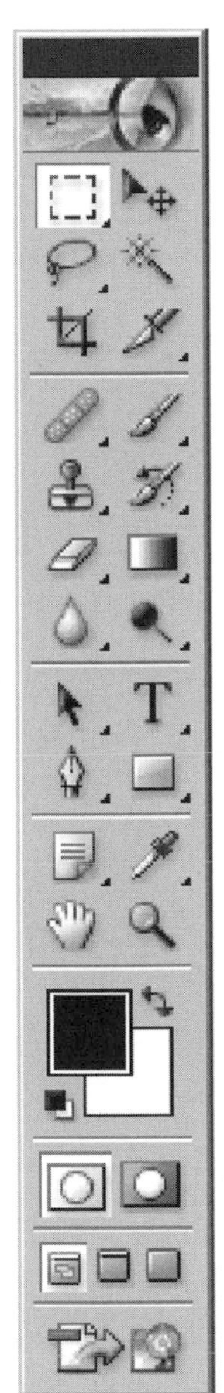

laptops, or mobile media devices (mobile phones, cameras, music players); and they are also likely to be unfamiliar with the details of twentieth-century cel animation, film editing equipment, or any other pre-digital media technology. Does this mean that the distinction between software simulations of previously existing media tools and new "born digital" media techniques has no meaning for digital natives but only matters for historians of media such as myself?

I think that while the semantics of this distinction (i.e. the reference to previous technologies and practices) may not be meaningful to digital natives, the distinction itself is something these users experience in practice. To understand why this is the case, let us ask if all "born digital" media techniques available in media authoring software applications may have something in common—besides the fact that they did not exist before software.

One of the key uses of digital computers from the start was *automation*. As long as a process can be defined as a finite set of simple steps (i.e. as an algorithm), a computer can be programmed to execute these steps without human input. In the case of application software, the execution of any command involves "low-level" automation (since a computer automatically executes a sequence of steps of the algorithm behind the command). However, what it is important from the user's point of view is the level of automation being offered in the command's interface.

Many software techniques that simulate physical tools share a fundamental property with these tools: they require a user to control them "manually." The user has to micro-manage the tool, so to speak, directing it step-by-step to produce the desired effect. For instance, you have to explicitly move the cursor in a desired pattern to produce a particular brushstroke using a brush tool; you also have to explicitly type every letter on a keyboard to produce a desired sentence. In contrast, many of the techniques that do not simulate anything that existed previously—at least, not in any obvious way—offer higher-level automation of creative processes. Rather than controlling every detail, a user specifies parameters and controls and sets the tool in motion. All generative (also called "procedural") techniques available in media software fall into this category. For example, rather than having to create a rectangular grid made of thousands of lines by hand, a user can specify the width and the height of the grid and the size of one cell, and the program will generate the desired result. Another example of this higher-level automation is interpolation of key values offered by animation software. In a twentieth-century animation production, a key animator drew key frames which were then forwarded to human in-betweeners who created all the frames *in between* the key frames. Animation software automates the process of creating in-between drawings by automatically interpolating the values between the key frames.

Thus, although users may not care that one software tool does something that was not possible before digital computers while another tool simulates previous physical or electronic media, the distinction itself between the two types is something users experience in practice. The tools that belong to the first type showcase the ability of computers to automate processes; the tools that belong to the second type use invisible low-level automation behind the scenes while requiring users to direct them manually.

Filter > stylize > wind

Having established two sets of categories for software techniques (media-independent *vs.* media-specific; simulation of the old *vs.* new), let us now test them against the Photoshop commands. Before starting, however, it is important to note once again that the two proposed schemes are intended to serve only as provisional categories. They provide one possible set of directions—an equivalent of North, South, West and East for a map where we can locate multiple operations of media design software. Like any first sketch, no matter how imprecise, this map is useful because now we have something to modify as we go forward. Our goal is not to try to fit everything we will look at into the categories of this initial map, but rather to discover its limitations as quickly as we can, and make modifications.

We will start with Photoshop filters—i.e. the set of commands that appear under the Filter menu. (Note that a large proportion of Photoshop filters are not unique to this program but are also available in other professional image editing, video editing and animation software—sometimes under different names. To avoid any possible misunderstanding, I will be referring to the Photoshop versions of these commands as implemented in Photoshop CS4, with their particular options and controls as defined in this software release.[14])

The first thing that is easy to notice is that the names of many Photoshop filters refer to the techniques for image manipulation and creation and materials that were available before the development of media application software in the 1990s—painting, drawing and sketching, photography, glass, neon, photocopying. Each filter is given a set of explicit options that can be controlled with interactive sliders and/or by directly entering desired numbers. These controls not only offer many options but also often allow you to set filter's properties numerically by choosing a precise value from a range.

This is a good example of my earlier point that simulations of prior physical media augment them with new properties. In this case, the new property is the explicit filter controls. For example, the Palette Knife filter offers three options: Stroke Size, Stroke Detail, and Softness. Stroke Size can take values between 1 and 50; the other two options have similarly large ranges. (At the same time, it is important to note that expert users of many physical tools such as a paintbrush can also achieve many effects not possible in its software simulation. Thus, software simulations should not be thought of simply as improvements over previous media technologies.)

While some of these filters can be directly traced to previous physical and mechanical media such as oil painting and photography, others make a reference to actions or phenomena in the physical world that at first appear to have nothing to do with media. For instance, the Extrude filter generates a 3D set of blocks or pyramids and paints image parts on their faces, while the Wave filter creates the effect of ripples on the surface of an image.

However, if we examine any of these filters in detail, we realize that things are not so simple. Let us take the Wind filter (located under the Stylize submenu) as an example. This is how Photoshop CS4's built-in Help describes this filter: "Places tiny horizontal lines in the image to create a windblown effect. Methods include Wind; Blast, for a more dramatic wind effect; and Stagger, which offsets the lines in the image." We are all familiar with the visual effects

[14] For a history of Photoshop version releases, see http://en.wikipedia.org/wiki/Adobe_Photoshop_release_history

The effects of Photoshop Wind filter (version CS5.1). Left to right: original shape, no filter; the filter with "wind" option applied; the filter with "blast" option applied.

of a strong wind on a physical environment (for instance, blowing through a tree or a field of grass)—but before you encountered this filter, you probably never imagined that you can "wind" an image. Shall we understand the name of this filter as a metaphor? Or perhaps, we can think of it as an example of a conceptual "blend" (which is how, according to Conceptual Blending theory, many concepts in natural languages get formed[15]): "wind" plus "image" results in a new concept actualized in the operations of the Wind filter.

The situation is further complicated by the fact that the results of applying the Wind filter to an image look pretty different compared to what the actual wind does to a tree or a field of grass. However, they do look rather similar to a photograph of a real windy scene taken with a long exposure. Therefore, we can think of the name "Wind" both as a metaphor—to help us imagine what a particular algorithmic transformation does to an image—and as a simulation of a particular photographic technique (long exposure). In short, although its name points to the physical world, its actual operations may also refer to a pre-digital media technology.

Are there "born digital" filters?

Let us continue the exploration of Photoshop filters. The great majority of the filters make references to previous physical media

[15] See Mark Turner and Gilles Fauconnier, *The Way We Think. Conceptual Blending and the Mind's Hidden Complexities* (New York: Basic Books 2002).

or our experiences in the physical world—at least in terms of how they are named. Only a few do not. These filters include High-pass, Median, Reduce Noise, Sharpen, and Equalize. Are these filters "born digital"? In other words, did we finally get to pure examples of "new" media techniques? The answer is no. As it turns out, all these filters are also software simulations that refer to things that already existed before digital computers.

Although they represent a small subset of Photoshop's extensive filter collection, these filters are central to all electronics, telecommunication, and IT technologies. Moreover, they are not unique to processing digital images but can be used on any kind of data—sounds, television transmission, data captured by an environmental sensor, data captured by a medical imaging devices, etc.

In their Photoshop implementation, these filters work on continuous-tone images, but since they can be also applied to sound and other types of signals, they actually belong to our "media-independent" category of software techniques. In other words, they are general techniques developed first in engineering and later also in computer science for signal and information processing. The application of these techniques to images forms the part of the field of image processing defined as "any form of information processing for which the input is an image, such as photographs or frames of video."[16] This conceptual relationship between "information processing" and "image processing" exemplifies one of the key points of this book—in software culture, "digital media" is a particular subset of the larger category "information." (Thus, the operations commonly used with media form a subset of the larger category "data processing.")

Like these filters, *many of the "new" techniques for media creation, editing, and analysis implemented in software applications were not developed specifically to work with media data.* Rather, they were created for signal and information processing in general—and then were either directly carried over to, or adapted to work with media. (Thus, development of software brings different media types closer together because the same techniques can be used on all of them. At the same time, "media" now share

[16] *Ibid.*

a relationship with all other information types, be they financial data, patient records, results of a scientific experiments, etc.)

This is one of the most important theoretical dimensions in the shift from physical and mechanical media technologies to electronic media and then digital software. Previously, physical and mechanical media tools were used to create content which was directly accessible to human senses (with some notable exceptions like Morse code)—and therefore the possibilities of each tool and material were driven by what was meaningful to a particular human sense. A paintbrush could create brushstrokes that had color, thickness, and shape—properties directly speaking to human vision and touch. Similarly, the settings of photographic camera controls affected the sharpness and contrast of captured photos—characteristics meaningful to human vision. A different way to express this is to say that the "message" was not encoded in any way; it was created, stored, and accessed in its native form. So if we were to redraw the famous diagram of a communication system by Claude Shannon (1948)[17] for the pre-electronics era, we would have to delete the encoding and decoding stages.

Successive media technologies based on electronics (such as the telegraph, telephone, radio, television), and digital computers employ the coding of messages or "content." And this, in turn, makes possible the idea of *information*—a disembodied, abstract and universal dimension of any message separate from its content. Rather than operating on sounds, images, video, or texts directly, electronic and digital devices operate on the continuous electronic signals or discrete numerical data. This allows for the definition of various operations that work on any signal *or any set of numbers*—regardless of what this signal or numbers may represent (images, video, student records, financial data, etc.). Examples of such operations are modulation, smoothing (i.e., reducing the differences in the data), and sharpening (exaggerating the differences). If the data is discrete, this allows for various additional operations such as sorting and searching.

The introduction of the coding stage allows for a new level of efficiency and speed in processing, transmitting, and interacting with media data and communication content—and this is why first

---

[17] C. E. Shannon, "A Mathematical Theory of Communication," *Bell System Technical Journal* 27 (July 1948): 379–423; (October 1948): pp. 623–56.

electronics and later digital computers gradually replaced all other media-specific tools and machines. Operations such as those just mentioned are now used to automatically process the signals and data in various ways—reducing the size of storage or bandwidth needed, improving quality of a signal to get rid of noise, and of course—perhaps most importantly—to send media data over communication networks.

The field of digital image processing began to develop in the second half of the 1950s when scientists and the military realized that digital computers could be used to automatically analyze and improve the quality of aerial and satellite imagery collected for military reconnaissance and space research. (Other early applications included character recognition and wire-photo standards conversion.[18]) As a part of its development, the field took the basic filters that were already commonly used in electronics and adapted them to work with digital images. The Photoshop filters that automatically enhance image appearance (for instance, by boosting contrast, or by reducing noise) come directly from that period (late 1950s to early 1960s).

In summary, Photoshop's seemingly "born digital" (or "software-native") filters have direct physical predecessors in analog filters. These analog filters were first implemented by the inventors of telephone, radio, telephone, electronic music instruments, and various other electronic media technologies during the first half of the twentieth century. They were already widely used in the electronics industry and studied in the field of analog signal processing before they were adapted for digital image processing.

Filter > distort > wave

The challenges in deciding in what category to place Photoshop filters persist as we continue going through the Filter menu. The two schemes for classifying software techniques for working with

[18] Even though image processing represents an active area of computer science, and is widely used in contemporary societies, I am not aware of any books or even articles that trace its history. The first book on image processing was Azriel Rosenfeld, *Picture Processing by Computer* (New York: Academic Press, 1969).

media I proposed turn out to be exactly what I suggested them to be—only an initial rough map to start a discussion.

The difficulties in deciding where to place this or that technique are directly related to the history of digital computers as simulation machines. Every element of computational media comes from some place outside of digital computers. This is true not only for a significant portion of media editing techniques—filters, digital paintbrushes and pencils, CAD tools, virtual musical instruments and keyboards, etc.—but also for the most basic computer operations such as sort and search, or basic ways to organize data such as a file or a database. Each of these operations and structures can, both conceptually and historically, be traced to previous physical or mechanical operations and to strategies of data, knowledge and memory management that were already in place before the 1940s. For example, computer "files" and "folders" refer to their paper predecessors already standard in every office. The first commercial digital computers from IBM were marketed as faster equivalents of electro-mechanical calculators, tabulators, sorters and other office equipment for data processing that IBM had already been selling for decades. However, as we already discussed in detail using as examples the work of Ivan Sutherland, and Douglas Engelbart's and Alan Kay's labs, whenever some physical operations and structures were simulated in a computer, they were simultaneously enhanced and augmented. This process of transferring physical world properties into a computer while augmenting them continues today—think, for instance of the multi-touch interface popularized by Apple's iPhone (2007). Thus, while Alan Turing defined the digital computer as a general-purpose simulation machine, when we consider its subsequent development and use, it is more appropriate to think of *a computer as a simulation-augmentation machine*. The difficulty of deciding how to classify different media software techniques is a direct result of this paradigm that underlies the development of what we now call software applications from the very start (i.e. Sutherland's Sketchpad, 1962–3).

The shift from physical media tools and materials as algorithms designed to simulate the effects of these tools and materials also has another important consequence. As we saw, some Photoshop filters explicitly refer to previous artistic media; others make reference to diverse physical actions, effects, and objects (Twirl, Extrude, Wind, Diffuse Glow, Open Ripple, Glass, Wave, Grain,

Patchwork, Pinch, and others). But in both cases, by changing the values of the controls provided by each filter, we can vary its visual effect significantly on the familiar/unfamiliar dimension. We can use the same filter to achieve a look that may indeed appear to closely simulate the effect of the corresponding physical tool or physical phenomena—or a look which is completely different from both nature and older media and which can be only achieved though algorithmic manipulation of the pixels. What begins as a reference to a physical world outside of the computer if we use default settings can turn into something totally alien with a change in the value of a single parameter. In other words, many algorithms only simulate the effects of physical tools and machines, materials or physical world phenomena when used with particular parameter settings; when these settings are changed, they no longer function as simulations.

For an example, let us analyze the behavior of the Wave filter (located under the Distort submenu). The filter refers to a familiar physical phenomena, and indeed, it can produce visual effects which we would confidently call "waves." This does not mean that the effect of this filter has to closely resemble the literal meaning of wave defined by a dictionary as "a disturbance on the surface of a liquid body, as the sea or a lake, in the form of a moving ridge or swell."[19] In our everyday language, we use the word "wave" metaphorically to refer to any kind of periodical movement ("waving a hand"), or any static form that resembles the form of a wave, or a disturbance in the ordinary state of affairs ("making waves.") According to an influential theory developed by cognitive linguist George Lakoff, such metaphorical use is not an exception, but the norm in human language and thinking. Lakoff proposed that the majority of our abstract concepts are metaphorical projections from sensorimotorial experiences with our own body and the surrounding physical world.[20] "Making waves" and other metaphors derived from our perceptual experience of seeing real waves exemplify this general mechanism of language.

If we follow Lakoff' s theory of metaphor, some details of the Wave filter operation—along with many other Photoshop filters

---

[19] http://dictionary.reference.com/browse/wave (August 1, 2012).

[20] George Lakoff and Mark Johnson, *Metaphors We Live By* (Chicago: University of Chicago Press, 1980).

Photoshop's Wave filter (version CS5.1). From left to right: the original shape, no filter; the filter with 1 wave generator; the filter with 10 wave generators; the filter with 50 wave generators.

that refer to the physical world—can be understood as similar metaphorical projections. Depending on the choice of parameter values, this filter can either produce effects that closely resemble our perceptual experience of actual physical waves, or new effects that are related to such waves metaphorically.

The filter generates sine wave functions (y = sin x), adds them up and uses the result to distort an image. A user can control the number of sine waves via a parameter called Number of Generators. If this number is set to 1, the filter generates a single sine wave. Applying this single function to an image distorts it using a periodically varying pattern. In other words, the filter indeed generates an effect that looks like a wave.

However, at some point the metaphorical connection to real world waves breaks, and using Lakoff's theory no longer works. If we increase the number of generators (it can go up to 999), the pattern produced by the filter no longer appears to be periodical, and therefore it no longer can be related to real waves even metaphorically.

The reason for this filter behavior lies in its implementation. As we have already explained, when the number of generators is set to 1, the algorithm generates a single sine function. If the option is to set to 2, the algorithm generates two functions; if it is set to 3, it generates three functions, and so on. The parameters of each function are selected randomly within the user specified range.

If we keep the number of generators small (2–5), sometimes these random values add up to a result that still resembles a wave; in other cases they do not. But when the number of functions is increased, the result of adding these separate functions with unique random parameters never looks like a wave.

Wave filters can create a practically endless variety of abstract patterns—and most of them are not periodic in an obvious way, i.e. they are no longer visually recognizable as "waves." What they are is the result of a computer algorithm that uses mathematical formulas and operations (generating and adding sine functions) to create a vast space of visual possibilities. So although the filter is called "Wave," only a tiny part of its space of possible patterns corresponds to the wave-like visual effects in the real world.

The same considerations apply to many other Photoshop filters that make references to physical media. Similar to the Wave filter, the filters gathered under the Artistic and Texture submenus produce very precise simulations of the visual effects of physical media with a particular range of parameter settings; but when the parameters are outside of this range, these filters generate a variety of abstract patterns.

The operation of these Photoshop filters has important theoretical consequences. Earlier I pointed out that the software tools that simulate physical instruments—paint brushes, pens, rulers, erasers, etc.—require manual control, while the tools that do not refer to any previous media offer higher level automation. A user sets the parameter values and the algorithm automatically creates the desired result.

The same high-level automation underlies "generative" (or "procedural") software techniques commonly used today. This work generated by algorithms ranges from live visuals and animations by software artist Lia[21] to the massive procedurally generated world in the videogame Minecraft. Other generative projects use algorithms to automatically create complex shapes, animations, spatial forms, music, architectural plans, etc. (A good selection of interactive generative works and generative animations can be found at http://www.processing.org/exhibition/). Since most artworks created with generative algorithms are abstract, artists and theorists like to oppose them to software such as Photoshop and Painter that are widely used by commercial illustrators and photographers in the service of realism and figuration. Additionally, because these applications simulate older manual models of creation, they are also seen as less "new media-specific" than generative software. The

---

[21] http://www.liaworks.com/category/theprojects/

end result of both of these critiques is that software that simulates "old media" are thought to be conservative, while generative algorithms and artworks are presented as progressive because these are unique to "new media." (When people claim that artworks that involve writing computer code qualify as "digital art" while artworks created using Photoshop or other media applications do not, they rehearse a version of the same argument.)

However, as they are implemented in media applications such as Photoshop, the software techniques that simulate previous media and the software techniques that are explicitly procedural and use higher-level automation are part of the same continuum. As we saw with the Wave filter, the same algorithm can generate an abstract image or a realistic one. Similarly, particle systems algorithms are used by digital artists and motion graphics designers to generate abstract animations; the same algorithms are also widely used in film production to generate realistic-looking explosions, fireworks, flocks of birds and other physical natural phenomena. In another example, procedural techniques often used in architectural design to create abstract spatial structures are also used in video games to generate realistic 3D environments.

History and actions menus

I started this discussion of Photoshop filters to test the usefulness of two schemes for classifying the seemingly endless variety of techniques available in media software: 1) media-independent *vs.* media-specific techniques (first scheme); 2) the simulations of previous tools *vs.* techniques which do not explicitly simulate prior media (second scheme). The first scheme draws our attention to the fact that all media applications share some genes, so to speak, while also providing some techniques that can only work on particular data types. The second scheme is useful if we want to understand the software techniques in terms of their genealogy and their relation to previous physical, mechanical, and electronic media.

Although the previous discussion highlighted the difficult borderline cases, in other cases the divisions are clear. For example, the Brush Strokes filter family in Photoshop clearly takes inspiration from earlier physical media tools, while Add Noise does not.

The Copy and Paste commands are examples of media-independent techniques; Auto Contrast and Replace Color commands are examples of media-specific techniques.

However, beyond these distinctions suggested by the two schemes I proposed, all software techniques for media creation, editing, and interaction also share some additional common traits that we have not discussed yet. Conceptually, these traits are different from common media-independent techniques such as copy and paste. What are they?

Regardless of whether they refer to some pre-existing instrument, action, or phenomena in the physical world or not, media techniques available in application software are implemented as computer programs and functions. Consequently, they follow the principles of modern software engineering in general. Their interfaces use established conventions employed in all application software—regardless of whether these tools are part of spreadsheet software, inventory management software, financial analysis software, or web design software. The techniques are given extensive numerical controls; their settings can be saved and retrieved later; their use is recorded in a History window so it can be recalled later; they can be used automatically by recording and playing Actions; and so on. (The terms "History palette" and "Actions" refer to Photoshop, but the concepts behind them are found in many other software applications.) In other words, they acquire the full functionality of the modern software environment—functionality that is significantly different from that of physical tools and machines that existed previously. Because of these shared implementation principles, all software applications are like species that belong to the same evolutionary family, with media software occupying a branch of the tree.[22]

The pioneers of media software aimed to extend the properties of media technologies and tools they were simulating in a computer—in each case, as formulated by Kay and Goldberg, the goal was to create "a new medium with new properties." Consequently, software techniques that refer to previous physical, mechanical, or electronic tools and creative processes are also "new media"

---

[22] Contemporary biology no longer uses the idea of an evolutionary tree; the "species" concept has similarly proved to be problematic. Here I am using these terms only metaphorically.

because they behave so differently from their predecessors. We now have an additional reason to support this conclusion. New functionality (for instance, multiple zoom levels), the presence of media-independent techniques (copy, paste, search, etc.) and standard interface conventions (such as numerical controls for every tool, preview option, or commands history) further separate even the most "realistic" media simulation tool from its predecessor.

This means that to use any media authoring and editing software is to use "new media." Or, to unfold this statement: *all media techniques and tools available in software applications are "new media"—regardless of whether a particular technique or program refers to previous media, physical phenomena, or a common task that existed before it was turned into software.* To write using Microsoft Word is to use new media. To take pictures with a digital camera is to use new media. To apply the Photoshop Clouds filter (Filters > Render > Clouds) that uses a purely automatic algorithmic process to create a cloud-like texture is to use new media. To draw brushstrokes using the Photoshop brush tool is to use new media.

In other words, regardless of where a particular technique would fall in our classification schemes, all these techniques are instances of one type of technology—interactive application software. And, as Kay and Goldberg explained in their 1977 article quoted earlier, interactive software is qualitatively different from all previous media. Over the next thirty years, these differences became only larger. Interactivity; customization; the possibility to both simulate other media and information technologies and to define new ones; processing of vast amounts of information in real-time; control and interaction with other machines such as sensors; support of both distributed asynchronous and real-time collaboration—these and many other functionalities enabled by modern software (of course, working together with middleware, hardware, and networks) separate software from all previous media and information technologies and tools invented by humans.

Layers palette

For our final analysis, we will go outside the Filter menu and examine one of the key features of Photoshop that originally differentiated it from many "consumer" media editors—the Layers palette. The Layers feature was added to Photoshop 3.0, released in 1994.[23] To quote Photoshop Help, "Layers allow you to work on one element of an image without disturbing the others."[24] From the point of view of media theory, however, the Layers feature is much more than that. It redefines both how images are created and what an "image" actually means. What used to be an indivisible whole becomes a composite of separate parts. This is both a theoretical point, and the reality of professional design and image editing in our software society. Any professional design created in Photoshop is likely to use multiple layers (in Photoshop CS4, a single image can have thousands of layers). Since each layer can always be made invisible, layers can also act as containers for elements that potentially may go into the composition; they can also hold different versions of these elements. A designer can control the transparency of each layer, group them together, change their order, etc.

Layers change how a designer or an illustrator thinks about images. Instead of working on a single design with each change immediately (and in the case of physical media such as paint or ink, irreversibly) affecting this image, s/he now works with a collection of separate elements. S/he can play with these elements, deleting, creating, importing and modifying them, until s/he is satisfied with the final composition—or a set of possible compositions that can be defined using Layer Groups. And since the contents and the settings of all layers are saved in an image file, s/he can always come back to this image to generate new versions or to use its elements in new compositions.

The layers can also have other functions. To again quote Photoshop CS4's online Help, "Sometimes layers don't contain any apparent content. For example, an adjustment layer holds color or tonal adjustments that affect the layers below it. Rather than edit image pixels directly, you can edit an adjustment layer and leave the underlying pixels unchanged."[25] In other words, the layers may contain editing operations that can be turned on and off, and re-arranged in any order. An image is thus redefined as a provisional composite of both content elements and various

---

[23] http://en.wikipedia.org/wiki/Adobe_Photoshop_release_history

[24] http://help.adobe.com/en_US/Photoshop/11.0/ (October 9, 2011).

[25] *Ibid.*

modification operations that are conceptually separate from these elements.

We can compare this fundamental change in the concept and practice of image creation with a similar change that took place in mapping—a shift from paper maps to GIS. Just as all media professionals use Photoshop, today the majority of professional users who deal with physical spaces—city offices, utility companies, oil companies, marketers, hospital emergency teams, geologists and oceanographers, military and security agencies, police, etc.—use GIS systems. Consumer mapping software such as Google Maps, Microsoft Bing Maps and Google Earth can be thought of as very simplified GIS systems. They do not offer the features that are crucial for professionals such as spatial analysis. (An example of spatial analysis is directing software to automatically determine the best positions for new supermarkets based on existing demographic, travel, and retail data.)

GIS "captures, stores, analyzes, manages, and presents data that is linked to location."[26] The central concept of GIS is a stack of data layers united by common spatial coordinates. There is an obvious conceptual connection to the use of layers in Photoshop and other media software applications—however, GIS systems work with any data that has geospatial coordinates rather than only images positioned on separate layers. The geospatial coordinates align different data sets together. As used by professionals, "maps" constructed with GIS software may contain hundreds or even thousands of layers. The layers representation is also used in consumer applications such as Google Earth. However, while in professional applications such as ArcGIS users can create their own layered maps from any data sources, in Google Earth users can only add their own data to the base representation of Earth that is provided by Google and cannot be modified.

In the GIS paradigm, space functions as a *media platform* which can hold all kinds of data types together—points, 2D outlines, maps, images, video, numerical data, text, links, etc. (Other types of such media platforms commonly used today are databases, web pages, and spaces created via 3D compositing that I will discuss later in the book). In Photoshop the layers are still conceptually

[26] http://en.wikipedia.org/wiki/GIS (October 9, 2011).

subordinated to the final image—when you are using the application, it continuously renders all visible layers together to show this image. So although you can use a Photoshop image as a kind of media database—a way to collect together different image elements—this is not the intended use (you are supposed to use separate programs such as Adobe Bridge or Aperture to do that). GIS takes the idea of a layered representation further. When professional users work with GIS applications, they may never output a single map that would contain all the data. Instead, users select the data they need to work with at that moment and then perform various operations on this data (practically, this means selecting a subset of all data layers available). If a traditional map offers a fixed representation, GIS, as its name implies, is an information system: a way to manage and work with a large sets of separate data entities linked together—in this case, via a shared coordinate system.

From programming techniques to digital compositing

What is the conceptual origin of Layers in Photoshop? Where do Layers belong in relation to my taxonomies of software-based media techniques? Thinking about various possible sources of this concept and also considering how it relates to other modern media editing techniques takes us in a number of different directions. First of all, layers are not specific to raster image editors such as Photoshop; this technique is also used in vector image editors (Illustrator), motion graphics and compositing software (After Effects), video editors (Final Cut), and sound editors (Pro Tools). In programs that work with time-based data—sound editors, animation and compositing programs, and video and film editors—layers are usually referred to as "channels" or "tracks"; these different terms point to particular physical and electronic media which a corresponding digital application has replaced (analog video switchers, multitrack audio recorders). Despite the difference in terms, the technique functions in the same way in all these applications: a final composition is a result of a "adding up" data (technically, a composite) stored in different layers/channels/tracks.

Photoshop Help explains Layers in the following way: "Photoshop layers are like sheets of stacked acetate. You can see through transparent areas of a layer to the layers below." (Photoshop CS4 Help, "About layers."[27]) Although not explicitly named by this Help article, the reference here is to the standard technique of twentieth-century commercial. Like a film camera mounted above the animation stand, Photoshop software is "shooting" the image created through a juxtaposition of visual elements contained on separate layers.

It is not surprising that Photoshop Layers are closely related to twentieth-century visual media techniques such as cel animation, as well as to various practices of pre-digital compositing such as multiple exposure, background projection, mattes in filmmaking, and video keying.[28] However, there is also a strong conceptual link between image Layers and twentieth-century music technology. The use of layers in media software to separate different elements of a visual and/or temporal composition strongly parallels the earlier practice of multitrack audio recording. The inventor of multitrack recording was the guitarist Les Paul; in 1953 he commissioned Ampex to build the first eight-track recorder. In the 1960s multi-track recorders were already being used by Frank Zappa, the Beach Boys, and the Beatles; from that point on, multitrack recording became the standard practice for all music recording and arranging.[29] Originally a bulky and very expensive machine, a multi-track recorder was eventually simulated in software and is now available in many applications. For instance, since 2004 Apple has included the multitrack recorder and editor GarageBand on all its new computers. Other popular software implementations include the free application Audacity and the professional-level application Pro Tools.

Finally, yet another lead links Layers to a general principle of modern computer programming. In 1984, two computer scientists Thomas Porter and Thomas Duff working for ILM (Industrial Light and Magic, a special effects unit of Lucasfilm) formally

---

[27] http://help.adobe.com/en_US/Photoshop/11.0/ (October 9, 2011).

[28] The chapter "Compositing" in *The Language of New Media* presents an "archeology" of digital compositing that discusses the links between these earlier technologies. Lev Manovich, *The Language of New Media* (The MIT Press, 2001.)

[29] See http://en.wikipedia.org/wiki/Multitrack_tape_recorder and http://en.wikipedia.org/wiki/History_of_multitrack_recording

defined the concept of *digital compositing* in a paper presented at SIGGRAPH.[30] The concept emerged from the work ILM was doing on special effects scenes for 1982's *Star Trek II: The Wrath of Khan*. The key idea was to render each separate element with a matte channel containing transparency information. This allowed the filmmakers to create each element separately and then later combine them into a photorealistic 3D scene.

Porter and Duff's paper makes an analogy between creating a final scene by compositing 3D elements and assembling separate code modules into a complete computer program. As Porter and Duff explain, the experience of writing software in this way led them to consider using the same strategy for making images and animations. In both cases, the parts can be re-used to make new wholes:

> Experience has taught us to break down large bodies of source code into separate modules in order to save compilation time. An error in one routine forces only the recompilation of its module and the relatively quick reloading of the entire program. Similarly, small errors in coloration or design in one object should not force "recompilation" of the entire image.[31]

The same idea of treating an image as a collection of elements that can be changed independently and re-assembled into new images is behind Photoshop Layers. Importantly, Photoshop was developed at the same place where the principles of digital compositing were defined earlier. Brothers Thomas and John Knoll wrote the first version of the program when Thomas took a six-month leave from the PhD program at the University of Michigan in 1988 to join his brother who was then working at ILM.

This link between a popular software technique for image editing and a general principle of modern computer programming is very telling. It is a perfect example of how all elements of the modern media software ecosystem—applications, file formats, interfaces, techniques, tools and algorithms used to create, view, edit, and share media content—have not just one but two parents,

---

[30] Thomas Porter and Tom Duff, "Compositing Digital Images," *Computer Graphics*, vol. 18, no. 3 (July 1984): p. 253–9.
[31] *Ibid.*

each with their own set of DNA: media and cultural practices on the one hand, and software development on the other.

In short, through the work of many people, from Ivan Sutherland in early 1960s, to the teams at ILM, Macromedia, Adobe, Apple and other companies in the 1980s and 1990s, *media becomes software*—with all the theoretical and practical consequences such a transition entails. This section dives into Photoshop's Filter and Layers menus to discuss some of these consequences—but more still remain to be uncovered.

There is only software

What exactly is "new media" and how is it different from "old media"? Academics, new media artists, and journalists have been writing extensively about this question since the early 1990s. In many of these discussions, a single term came to stand for the whole range of new technologies, new expressive and communicative possibilities, new forms of community and sociality that were emerging around computers and Internet. The term is "digital." It received its official seal of approval, so to speak, in 1996 when the director of MIT Media Lab Nicholas Negroponte collected his *Wired* columns into the book that he named *Being Digital*.[32] Many years later, the term "digital" still dominates both popular and academic understanding of what new media is about.

When I did Google searches for "digital," "interactive," and "multimedia" on August 28, 2009, the first search returned 757 million results; the other two only returned 235 and 240 million respectively. Making searches on Google Scholar produced similar results: 10,800,000 for "digital," 4,150,000 for "web," 3,920,000 for "software," 2,760,000 for "interactive," 1,870,000 for "multimedia." Clearly, Negroponte was right: we have become digital.

I do not need to convince anybody today about the transformative effects the Internet, the web, and other technological networks have already had on human culture and society. However, what I do want to convince you of is the crucial role of another part of the computer revolution that has been less discussed. And yet, if we

[32] Nicholas Negroponte, *Being Digital* (Vintage, 1996).

really want to understand the forms of contemporary media and also what "media" means today, this part is crucial. The part in question is software.

None of the new media authoring and editing techniques we associate with computers are simply a result of media "being digital." The new ways of media access, distribution, analysis, generation, and manipulation all come from *software*. Which also means that they are the result of the particular choices made by individuals, companies, and consortiums who develop software—media authoring and editing applications, compression codecs, file formats, programming and scripting languages used to create interactive and dynamic media such as PHP and JavaScript. Some of these choices determine conventions and protocols which define modern software environments: for instance, "cut" and "paste" commands built into all software running under the Graphical User Interface and its newer versions (such as iPhone OS), or one-way hyperlinks as implemented in World Wide Web technology. Other choices are specific to particular types of software (for instance, illustration programs) or individual software packages.

If particular software techniques or interface metaphors which appear in one application—be it a desktop program, web application, or mobile app—become popular with its users, they may often soon appear in other apps. For example, after Flickr added tag clouds to its interface, they soon appeared on numerous other websites. The appearance of particular techniques in applications can also be traced to the economics of the software industry—for instance, when one software company buys another company, it may merge its existing package with the software from the company it bought. For instance, in 1995 Silicon Graphics bought two 3D computer graphics suites—Wavefront and Alias—and merged them into a new product called Alias|Wavefront. Big companies such as Google and Facebook are periodically buying smaller companies and then adding the software products these companies develop to their own offerings. Thus, one of Google's most popular applications, Google Earth is based on software originally developed by Keyhole, Inc. and acquired by Google in 2004.

Often, techniques created for one purpose later migrate into another area, as happened when image processing techniques made their way into Photoshop in the late 1980s. These techniques, developed in the second half of the 1950s for the analysis of reconnaissance photographs, are now used to creatively modify images and to make photographs more "artistic looking."

All these software mutations and new species of software techniques are deeply social—they do not simply come from individual minds or from some "essential" properties of a digital computer or a computer network. They come from software developed by groups of people, marketed to large numbers of users, and then constantly refined and expanded to stay competitive in relation to other products in the same market category. (Google and Facebook update their code a few times a day; GitHub, the popular software hosting services, updates its code dozens of times a day.)

In summary: the techniques, the tools, and the conventions of media software applications are not the result of a technological change from "analog" to "digital" media. The shift to digital enables the development of media software—but it does not constrain the directions in which it already evolved and continues to evolve. They are the result of intellectual ideas conceived by the pioneers working in larger labs, the actual products created by software companies and open source communities, the cultural and social processes set up when many people and companies start using it, and software market forces and constraints.

This means that the terms "digital media" and "new media" do not capture very well the uniqueness of the "digital revolution." (I like the term "media computing"—however it is not used widely apart from some communities in computer science primarily in Europe). Why do they not work? Because all the new qualities of "digital media" are not situated "inside" the media objects. Rather, they all exist "outside"—as commands and techniques of media viewers, authoring software, animation, compositing, and editing software, game engine software, wiki software, and all other software "species." While digital representation makes it possible for computers to work with images, text, 3D forms, sounds and other media types in principle, it is the software that determines what we can do with them. So while we are indeed *"being digital," the actual forms of this "being" come from software.*

Accepting the centrality of software puts in question another fundamental concept of aesthetic and media theory—that of the "properties of a medium." What does it mean to refer to a "digital medium" as having "properties"? For example, is it meaningful to talk about unique properties of digital photographs, electronic texts, or websites? In their article, Kay and Goldberg do pair the words "properties" and "medium" together: "It [electronic text] need not be treated as a simulated paper book since this is a new medium with new properties." I have also frequently used this combination of words—but it is now time to ask if it is only an alias that can point us to a more precise concept.

Strictly speaking, while it is certainly convenient to talk about properties of websites, digital images, 3D models, GIS representations, etc., it is not accurate. Different types of digital content do not have any properties by themselves. *What as users we experience as properties of media content comes from software used to create, edit, present, and access this content.*

This includes all media authoring and viewing application software made for both professionals and consumers, from Photoshop to your mobile web browser. (It also includes custom software developed for particular products such as a DVD menu or an interactive kiosk.) So while I will continue to use the term "properties" as a shortcut, you should always remember that it stands for *software techniques defined to work on particular types of media ecologies, content and media data.* (Flickr's whole system for uploading, tagging, organizing, commenting, and sharing images is an example of "media ecology"; a raster 24-bit image stored in JPEG format is an example of a type of "media data.")

It is important to make clear that I am not saying that today all the differences between media types—continuous tone images, vector images, simple text, formatted text, 3D models, animations, video, maps, music, etc.—are completely determined by application software. Obviously, these media data types have different representational and expressive capabilities; they can produce different emotional effects; they are processed by different sensors and networks of neurons in the brain; and they are likely to correspond to different types of mental processes and mental representations. These differences have been discussed for thousands of years—from ancient philosophy and classical aesthetic theory to modern art and contemporary neuroscience. What I am arguing in this book is something else. Firstly, *interactive software adds a new set of operations which can be applied to all media types—which we as users experience as their new "properties."* (The examples include the ability to display the same data structure in different ways, hyperlinking, visualization, searchability, and findability.) Secondly, *the "properties" of a particular media type can vary dramatically depending on the software application used for its authoring and access.*

Let us go though one example in detail. As an example of a media type, we will use a photograph. In the analog era, once a photograph was printed, all the information was "fixed." Looking at this photograph at home, in an exhibition, or in a book did not affect this information. Certainly, a photographer could produce a different print with a higher or a lower contrast or use a different paper—but this resulted in a physically different object, i.e., a new photographic print that contained different information. (For example, some details were lost if the contrast was increased.)

So what happens with a digital photograph? We can take a photo with a dedicated digital camera or capture it with a mobile phone, or scan it from an old book. In every case, we end up with a file that contains an array of pixels which hold color values, and a file header that specifies image dimensions, color profile, information about the camera and shot conditions such as exposure, and other metadata. In other words, we end up with what is normally called "digital media"—a file containing numbers which represent the details of some scene or an object.

However, unless you are a programmer, you never directly deal with these numbers. Instead, most of us interact with digital media files via some application software. And depending on which software you use, what you can do with a particular digital media file can change dramatically. MMS (multimedia messaging service) software on your phone may simply display a photo sent by a friend—and allow you to forward it to somebody else—but nothing else.

Free media viewers/players that run on desktops or on mobile platforms typically give you more functions. For instance, a desktop version of Google's Picasa 3.0 includes crop, auto color, red eye reduction, variety of filters (soft focus, glow, etc.) and a number of other functions. It can also display the same photo as color or black and white—without modifying the actual digital media file.

Finally, if I open the same photo in Photoshop, I can instruct Photoshop to automatically replace some colors in a photo with others, make visible its linear structure by running an edge detection filter, blur it in a dozen of different ways, composite with another photo, and perform hundreds of other operations.

To summarize this discussion, let me make a bold statement. *There is no such thing as "digital media." There is only software—as applied to media (or "content")*. Or, to put this differently: *for users who only interact with media content through application software, the "properties" of digital media are defined by the particular software as opposed to solely being contained in the actual content (i.e., inside digital files).*

"Digital media" is a result of the gradual development and accumulation of a large number of software techniques, algorithms, data structures, and interface conventions and metaphors. These techniques exist at different levels of generality ranging from a small number of very general ("media-independent") ones to thousands of very particular ones designed to do particular tasks—for example, algorithms used to generate natural-looking landscapes or software which can extract the camera position from live action footage in order to correctly align a 3D model when it is composited with this footage.

Because of the multiplicity and variety of these software techniques, it is unwise to try to reduce "digital media" to a small set of new properties. Such reduction would only be possible if we could organize all these techniques hierarchically, seeing them as different applications of a few general principles. After thinking on and off about this for ten years (starting with my 1999 article "Avant-Garde as Software" where I first tried to provide a taxonomy of these new techniques) I eventually came to the conclusion that such hierarchy will only mislead us. The reason is that all these techniques equally change the identity of any media type to which they are applied.

The fact that one technique may appear in many software packages designed to work with different media types (what I called "media-independent" techniques) while another technique may be specific to a particular type of media (I called these techniques "media-specific") does not make the latter any less theoretically important than the former. For instance, because a zoom function is present in word processors, media viewers, animation software, 3D modeling software, web browsers, etc., this does not make it more important than the algorithm designed to do only one particular thing in relation to one media type—for instance, a "spherize" command which modifies coordinates of all the points in a 3D polygonal model so it appears more spherical.

I do not think that we can qualitatively measure the practical effects on cultural production of both types of operations in this example to conclude that one is more radical than the other. Both operations change the media they act upon qualitatively, rather than quantitatively. They both add new qualities (or "affordances") to media which it did have before. A Word document which can be zoomed across multiple scales to reveal many pages at once has a different "media identity" from one which cannot. Similarly, the ability to precisely spherize a 3D model is a new way of working with a spatial form that did not exist before 3D software.

In *Avant-Garde as Software* I grouped all new techniques of digital media into four types based on what functions they support: access, generation, manipulation, and analysis. But even such simple differentiation appears problematic to me today—partly because of the evolution of software since 1999, which led to a gradual integration of these functions. For example, when a user selects a media file on his/her laptop, tablet or a phone, the file automatically opens in a media player/viewer program. And today most media viewers and players (Windows Media Player, Apple's QuickTime Player, etc.) offer some basic editing functions. Therefore, in practical terms today you cannot simply "access" media without automatically being offered some ways to "modify" it. (To be clear, I am talking here about personal computers and mobile devices and not specialized hardware specifically designed to offer only access, and to prevent modification of commercial digital content—such as DVD players or MP3 players.)

How did we arrive at this new situation where instead of looking at or reading content directly, most of us always experience it through the layer of applications? The seemingly obvious answer is the adoption of numerical code as the new universal intermediary. I call it intermediary because in order to make media accessible to our senses, it has to be analog—a travelling wave of oscillating pressure which we experience as sound, the voltage levels applied to the pixel elements of an LCD which makes them appear as different colors, different amounts of dyes deposited on paper by dye-sublimation printers, and so on. Such conversions from A to D (analog to digital) and D to A (digital to analog) are central for digital media functioning: for example, from light waves to numbers stored in a file representing the image, and then back to the voltage levels controlling the display. Or, in another example, when we design an object to be printed on a 3D printer, an analog representation on the screen is translated by a computer into a digital file that then drives the analog signals controlling the printer.

The two levels of encoding—first, a sampling of a continuous analog signal to create its representation using a scale of discrete numbers (for example, 256 levels commonly used to represent grey tones in images), followed by a translation of this discrete representation into a binary numerical system—make "media" incomprehensible for direct observation. The main reason for this is not the binary code *per se* (invented by the Indian scholar Pingala between the fifth and second century BC) since it is possible to learn how to convert in your head a binary notation into a decimal one. The problem is that representing even one image digitally requires lots of numbers. For example, an image in HD resolution (1920 × 1080) contains 2,073,600 pixels, or 6,220,800 distinct RGB values—making it very hard to comprehend the patterns such a set of numbers may represent if you examine them directly. (In passing: because of these considerations, any digital image can be understand as information visualization—revealing patterns contained in its numerical representation.)

Because looking at such sets of numbers with our bare eyes is meaningless, we need to employ some technologies to translate them into analog representations acceptable to our senses. Most often, an image file is translated by digital hardware and software into an image appearing on our screen. However, a digital representation of one type of media can also be translated into another media type that is meaningful to our senses. For example, in audio-visual performances software often uses video to drive sound, or reversely uses sound to generate abstract visuals. (Interestingly, the precursor to Edison's 1877 phonograph—the first device to record and reproduce sound—was Édouard-Léon Scott de Martinville's 1857 phonautograph that transcribed sound into a visual media. In other words, sound visualization was invented before sound recording and reproduction.)

From the beginning, technologies that generated and transmitted electro-magnetic analog signals (e.g, the gramophone) included at least some controls for its modification such as changing signal amplitude. The first well-known electronic instrument invented by Léon Theremin in 1920 turned such controls into a new paradigm for music performance. A performer controlled amplitude (volume) and frequency (pitch) of a sound by moving his/her hands closer or further away from the two antennas.

Software significantly extends this principle by including more controls and more ways of representing the data. For example, I can choose to display this text I am writing now in Word as an outline, or select a "Print Layout" which will show me boundaries of pages; I can choose to see footnotes or hide them; I can ask the application to automatically summarize the text; I can change different font families and sizes, and so on. Thus, while the actual data as it is represented and stored in a computer is no longer directly accessible to our senses, the new model of encoding and access has other significant advantages since the data can be formatted in a variety of ways. This formatting can be changed interactively; it can be also stored with the data and recalled later.

This discussion can help us to understand the relations between earlier electro-magnetic recording and reproduction technologies, which were developed in the last decades of the nineteenth century, and media software developed 100 years later. (Telephone: Bell, 1875; phonograph: Edison, 1878; television: Nipkow, 1884; radio: Fessenden, 1900.) While previous reproduction technologies such as woodblock printing, moveable type printing, printmaking, lithography, and photography retained the original form of media, the media technologies of the late nineteenth century abandoned it in favor of an electrical signal. In other words, they introduced *coding* as a way to store and transmit media. Simultaneously, these technologies also introduced a fundamentally new layer of media—*interface*, i.e. the ways to represent ("format") and control the signal. And this in its turn changes how media functions—its "properties" were no longer solely contained in the data but were now also depend on the interface provided by technology manufacturers.

The shift to digital data and media software 100 years later generalized this principle to all media. With all data types now encoded as sets of numbers, they can only be accessed by users via software applications which translate these numbers into sensory representations. The consequence of this is what we have already discussed: all "properties of digital media" are now defined by the particular software as opposed to solely being contained in the actual content, i.e. digital files. So what was already true for audio recording, radio, television, and video now also applies to text, images, and 3D.

In short: media becomes software.

I could conclude this chapter here—however we need to do one more thing. It would be unfair to direct all this attention to the term "digital media" without also taking up another term related to it. Today this term is both widely used and also often put into question. The term is "new media."

Since we now understand that "media" today is really a set of software techniques constantly in development, this gives a new meaning to this troubled term. Just as there is no logical limit to the number of algorithms which can be invented, people can always develop new software techniques for working with media. So from this perspective, the term "new media" captures well this fundamental logic of "the computer metamedium." *Software-based media will always be "new" as long as new techniques continue to be invented and added to those that already exist.* And while not every one of these techniques will change significantly how particular media or a combination of media will function, many will.

This logic of "permanent extendability" of media software follows the logic of software engineering as a whole. Computer scientists working in the academy and software companies constantly develop modifications of the existing algorithms, apply algorithms from one area in another area, and develop new algorithms. This logic—which can be also called "permanent innovation"—is different from the logic which governed the development of media technologies of the industrial age. Take cinema, for instance, from 1890 to 1990, (i.e. until the adoption of software tools by the film industry). Although the construction of film lenses, the properties and types of film stock, the operations of film camera, and other elements of film technology changed dramatically over the course of the twentieth century, the basic "representational capacity" of this image type remained the same. From Edison to George Lucas, film images continued to be produced in the same way: by focusing light reflected from the objects via lens on a flat plane. This type of capture process creates particular types of images of visible reality. Objects are rendered in one-point linear perspective. This means that the geometric properties of any scene regardless of its content are subjected to the same type of transformation—the same mapping that preserves some properties of the visible (perspective distortion related to distance of objects from the observer) as opposed to other properties (the real proportions of object sizes, for example). Additionally once a recording was made, the geometry of the image could not be modified later.

Once software enters the picture, these constants of the film medium become variables. While looking no less "real," film image can now have multiple relationships to the world being imaged. Digital compositing allows for the seamless insertion of 3D computer-generated models that were not present in the original scene. Conversely, the objects that were present can be seamlessly removed from images. Interactive virtual panoramas which allow the user to move around the space can be constructed automatically. In some cases, it is even possible to re-render a film sequence as though it was shot from a different point of view. And these are just some of the new ways in which new software changes film identity. (All these new possibilities, of course, also apply to video.)

PART TWO

Hybridization and evolution

CHAPTER THREE

Hybridization

Hybridity *vs.* multimedia

"The first metamedium" envisioned by Kay and Goldberg in 1977 has gradually become a reality in the 1990s. Beginning with Sketchpad and extending to the latest versions of media software, most physical media were simulated in detail and many new properties were added to them in the process. In parallel, a number of brand new computational media which have no physical precedents were also invented—for instance, interactive navigable 3D space (Ivan Sutherland), interactive multimedia (Architecture Machine Group's Aspen Movie Map); hypertext and hypermedia (Ted Nelson); interactive narrative film (Graham Weinbren); the Internet (Licklider, Bob Taylor, Larry Roberts, and others); the World Wide Web (Tim Berners-Lee); social media services (SixDegrees.com),[1] collaborative large-scale authoring platforms such as Wikipedia (Jimmy Wales and Larry Sanger), and so on.

New fundamental techniques for media generation, manipulation, and presentation which also had no previous physical equivalent were also developed—algorithmic generation of line images, 3D photorealistic rending, and the constraints originally introduced in Sketchpad. New media-specific and general (i.e.,

[1] D. M. Boyd and N. B. Ellison, "Social network sites: Definition, history, and scholarship," *Journal of Computer-Mediated Communication*, *13*(1) (2007), http://jcmc.indiana.edu/vol13/issue1/boyd.ellison.html

media-agnostic) data management techniques were introduced. Most importantly, by the middle of the 1990s computers became fast enough to "run" all these media. In short, Kay's vision of a computer as metamedium—a platform housing many existing and new media—was realized.

So what happens next? What is the next stage in the metamedium development? (I am using the word "stage" here in a logical rather than a historical sense). This is something that, as far as I can see, the inventors of computational media—Sutherland, Nelson, Engelbart, Kay and all the people who worked with them—did not write about. However, since they set up all the conditions for this next stage, they are indirectly responsible for it.

The discussion of the computer metamedium at the end of Kay and Goldberg's 1977 article creates the impression that it would develop via *additions*, as users built new types of media to suit their needs using the tools provided with the personal computer. Looking at the actual development of the computer metamedium over the following thirty years seems to confirm this conclusion.

For instance, if we look at the use of computer media in art which begins in the second half of the 1950s (music composition and algorithmic image generation), by 2003 the authoritative book *Digital Art* by Christine Paul already lists dozen of different areas. A Wikipedia article on "Collaborative software" similarly lists about a dozen program types (and of course there are dozens or hundreds of separate products for each type).[2] Another Wikipedia article on "Social Software"[3] lists twenty major types of social media (instant messaging, text chat, groupware, blogs, etc.)—and none of them existed in practice in the early computer days of the 1960s.

To continue with these examples of such additions, a typical visual design created today with software applications may also appear as a simple sum of previous media: for example, a pen drawing plus oil painting plus a photograph plus collage. Looking at the interfaces of media editing software seems to also confirm this impression. You see endless options for modifying a document, appearing one after another in multiple menus. As the new options

---

[2] http://en.wikipedia.org/wiki/Collaborative_software (February 4, 2012).

[3] http://en.wikipedia.org/wiki/Social_software (February 4, 2012).

become available—because software manufacture has released a new version, or you have purchased some plug-ins of your own—they appear as additions in the same menus. And certainly, the overall number of commands in popular media applications gradually increases over time from release to release—adding more techniques for authoring, editing, remixing, and creating outputs for different distribution platforms.

However, these processes of addition and accumulation are not the only ones defining evolution of the computer metamedium. While they are certainly at work, I think that they are not in the center of the transformation—or if you like, mutation—of this metamedium (and by extension, of all modern culture created via software) in the three decades following Kay and Goldberg's seminal article in 1977. (I use this year—of course, only symbolically—to mark the completion of the first "media invention" stage.)

I believe that the new period that began in the late 1970s represents a fundamentally distinct second stage in the development of a computer metamedium, a stage that follows the first stage of its invention and initial practical implementation. This new stage is *media hybridization*.

Once computers became a comfortable home for a large number of simulated and new media, it is only logical to expect that they would start creating hybrids. And this is exactly what has been taking place at this new stage in media evolution. Both the simulated and new media types—text, hypertext, still photographs, digital video, 2D animation, 3D animation, navigable 3D spaces, maps, location information and social software tools—came to function as building blocks for many new media combinations.

Here are some examples of computational *media hybrids*. Google Earth combines aerial photography, satellite imagery, 3D computer graphics, still photography, and other media to create a new hybrid representation which Google engineers called a "3D interface to the planet." A motion graphics sequence may combine content and techniques from different media such as live action video, 3D computer animation, 2D animation, painting, and drawing. (Motion graphics are animated visuals that surround us every day; the examples are film and television titles, TV graphics, the graphics for mobile media content, and non-narrative parts of commercials and music videos.) A website design may blend photos, typography, vector graphics, interactive elements, and animation. Physical installations integrated into cultural and commercial spaces—for example, Nobel Field at the Nobel Peace Center in Oslo by Small Design, interactive store displays for Nokia and Diesel by Nanika, or the interactive Memory Wall at Puerta America hotel in Madrid by Karen Finlay and Jason Bruges, and so on—combine animations, video, computer control, and various interfaces from sensors to touch, to create interactive spatial media environments.[4]

The built-in dictionary of Microsoft Word which I am using to write this text has a few definitions for "hybrid" including the following: "a plant produced from a cross between plants with different genetic constituents"; and "an animal that results from mating of parents from two distinct species or subspecies." These biological meanings of "hybrid" capture well what has been happening to media following its "softwarization" in the 1980s and 1990s—that is, the systematic translation of numerous techniques for media creation and editing from physical, mechanical, and electronic technologies into software tools. Translated into software, media techniques start acting like species within a common ecology—in this case, a shared software environment. Once "released" into this environment, they start interacting, mutating, and making hybrids.

If we want to relate the beginning of the new hybridization stage to some important projects and technologies in the history of computational media, the famous Aspen Movie Map interactive hypermedia system developed at MIT in 1978–9 would qualify as the starting point.[5] A precursor to Google Street View (launched in 2007), the system combined film of streets in Aspen, still photographs, a navigation map that featured both aerial photography and diagrammatic drawings, and audio. The second key event is the release of QuickTime multimedia software by Apple on December 2, 1991 as an addition to System Software 6. As Apple explained in "QuickTime 1.0: 'You Oughta be in Pictures" technical article (Summer 1991): "The recently introduced QuickTime 1.0 makes it easy for you to add dynamic media like video and sound into your

---

[4] http://www.davidsmall.com/articles/2006/06/01/nobel-field/; http://www.nanikawa.com/; http://www.jasonbruges.com/projects/international-projects/memory-wall

[5] "The Interactive Movie Map: A Surrogate Travel System," video, (The Architecture Machine, 1981), http://www.media.mit.edu/speech/videos/

applications – and that's just the beginning."[6] In the next ten years, commercial developers, engineers, designers, and independent media artists put lots of their creative energy into exploring the new ability of a computer to present multiple media. Thus, I think of the 1990s as the foundational period when many fundamental ways of combining media within the single computer platform were invented—followed by the next period of commercialization of these inventions in the 2000s (for example, Google Earth, introduced in 2005) and adoption to mobile platforms (iPhone, introduced in 2007, was able to play video and songs, display photos and hybrid maps, and send MMS).

To start the discussion of media hybridity, it is important to make it clear that I am not simply talking about something that already has a name—"computer multimedia," or simply "multimedia." This term became popular in the 1990s to describe applications and electronic documents in which different media types exist next to each other.

Often these media types—typically text, graphics, photographs, video, 3D scenes, and sound—are situated within what looks visually like a two-dimensional page. Thus a typical Web page of the 2000s is an example of multimedia; so is a typical PowerPoint presentation. Today, at least, this is the still the most common way of structuring multimedia documents. In fact, it is built into the workings of most media authoring application such as presentation software or web design software. When a user of Word, PowerPoint or Dreamweaver creates a "new document," s/he is presented with a white page ready to display typed text; other media types have to be "inserted" into this page via special commands. But interfaces for creating "multimedia" do not necessarily have to follow this convention. Email and multimedia messaging use another common paradigm for putting elements of different media types together—"attachments." Thus, a user of a mobile phone that supports MMS can send text messages with attachments that can include picture, sound, video and rich (formatted) text. Yet another paradigm persistent in digital culture—from Aspen Movie Map (1978) to VRML (1994–) to Second Life (2003–)—uses 3D space as the

[6] Apple, "QuickTime 1.0: 'You Oughta be in Pictures" (summer 1991), http://www.mactech.com/articles/develop/issue_07/Ortiz_Text.html

default platform with other media such as video attached to or directly inserted into this space.

"Multimedia" was an important term when interactive cultural applications, which featured a few media types, started to appear in large numbers in the early 1990s. The development of these applications was facilitated by the introduction of the appropriate storage media, i.e. recordable CD-ROMs in 1991, computer architectures and file formats designed to support multiple media file formats (Apple's QuickTime, 1991–) and multimedia authoring software (the first version of the VideoWorks software which later was renamed Macromedia Director was released in 1985). By the middle of the 1990s digital art exhibitions featured a variety of multimedia projects; digital art curricula began to offer courses in "multimedia narrative"; and art museums such as the Louvre started to publish multimedia CD-ROMs offering tours of their collections. In the second part of the decade multimedia took over the Web as more and more websites began to incorporate different types of media. By the end of the decade, "multimedia" became the default in interactive computer applications. Multimedia CD-ROMs, multimedia websites, interactive kiosks, and multimedia communication via mobile devices became so commonplace and taken for granted that the term lost its relevance. So while today we daily encounter and use computer multimedia, we no longer wonder at the amazing ability of computers and computer-enabled consumer electronics devices to show multiple media types at once.

Seen from the point of view of media history, "computer multimedia" is certainly a development of fundamental importance. Previously "multimedia documents" combining multiple media types—such as medieval illustrated manuscripts, sacred architecture, or twentieth-century cinema and television—were not interactive (in the sense of particular affordances provided by interactive computers, rather than other interactive technologies such as paper books) or networked. But co-existence of multiple media types within a single document or an application is only one of the new developments enabled by simulation of many media types in a computer. In putting forward the term *hybrid media* I want to draw attention to another, equally fundamental development that, in contrast to multimedia, so far has not received a formal name.

Certainly, it is possible to conceive of multimedia as a particular case of hybrid media. However, I prefer to think of them as overlapping but ultimately two different developments. While some classical multimedia applications of the 1990s would qualify as media hybrids, most will not. Conversely, although media hybrids often feature content in different media, this is only one aspect of their make-up. So what is the difference between the two? In multimedia documents and interactive applications, content types in multiple media appear *next to each other*. In a web page, images and video appear next to text; a blog post may similarly show text, followed by images and more text; a 3D world may contain a flat screen object used to display video. Alternatively, each element of a multimedia message opens in its own viewer (this was the case for MMS implementations in the phones of the 2000s). In contrast, in media hybrids, interfaces, techniques, and ultimately the most fundamental assumptions of different media forms and traditions, are brought together resulting in new *media gestalts*. That is, they merge together to offer a coherent new experience different from experiencing all the elements separately.

Another way to underline this difference is by using the metaphor of *sexual reproduction*. The result of sexual reproduction is new individuals that combine the genetic material of their parents—rather than just mechanical assemblages of parents' physical parts (which would be analogous to multimedia). Using this metaphor, we can say that new media offspring similarly combine the DNA of their media parents.

A related model that can help us to grasp some aspects of this process is that of *biological evolution*. This process results in new organisms, new species, and also new building blocks of the organisms (molecules such as DNA and proteins.) Similarly, sometimes new media offspring are only slightly different from the ones that already exist; at other times the combinations of software DNA produce distinct new media "species." The process of *media evolution* also produces new techniques for media authoring, editing, sharing, and collaborating, new interface conventions, and also new algorithms—the equivalents of the new building blocks of biological evolution.

Yet another metaphor that can help us to understand the new stage of the media development is *remix*. In the process of the computer metamedium development, different media types get remixed together, forming new combinations. Parts of these combinations enter into new remixes, ad infinitum.

Every metaphor highlights some aspects of a phenomenon while hiding or even distorting other aspects. The metaphors of sexual reproduction, biological evolution, and music remix work similarly. Each has advantages and disadvantages in explaining the second stage of computational media development. In this chapter, I will extensively use the concepts of hybridity and evolution to describe the new stage in media evolution. The chapter in Part 3 will work with the remix metaphor.

Since I will be invoking each of these metaphors in different parts of my narrative, you may get the impression that they are complementary parts of the same description. But this is not the case for hybridity and biological evolution meataphors. Besides the everyday meaning of "hybridity," it is also used in evolutionary theory in a particular way. So if we think of hybridity in that sense, we cannot use this concept and the biological evolution model at the same time.

Contemporary theories of biological evolution share a basic definition of species as a pool of organisms that can enter into sexual relations between themselves but not with other species. Through the processes of evolution most often predicated on geographical separation between groups of the same species, these groups change and eventually no longer can have reproductive relations with one another. Such groups become new species.

In contrast, an animal hybrid is the result of interbreeding between species. Most hybrids are produced artificially, although a few have been observed in nature.[7] Thus, hybrids are exceptions in the normal evolutionary process. Therefore, when I use the term "hybrid," I am relying on a more general meaning of this word outside of biology.

I also need to make a note about my use of the biological evolution model. I am not suggesting that computational media (or techno-cultural development in general) indeed "evolve" like biological mechanisms, and that the mechanisms of such evolution are the same as the mechanisms of biological evolution as formulated in contemporary biology. (In his 2007 book *Graphs, Maps, Trees: Abstract Models for Literary History*, Franco Moretti provides convincing explanations of why some of the key ideas

---

[7] M. L. Arnold. *Natural Hybridization and Evolution* (New York: Oxford University Press, 1996).

of biological evolution do not fit the cultural history.[8]) Instead, I want to use evolutionary theory as a rich conceptual toolbox, which can help us to think about any kind of *temporal process*. Understood in this way, evolutionary theory joins other theories of development that aim to explain physical, social, or psychological processes—each providing its own unique concepts which give us additional ways to conceptualize any development. Examples include Marx's theory of social development with its concepts of mode of production, base, superstructure, and so on; Freud's theory of "the dream work" as formulated in his 1899 *Interpretation of Dreams* (concepts of condensation and displacement); complex systems theory (concepts of emergence and self-organization); the phase change model from thermodynamics; and many others.

Having explained how I will use the concepts of hybridity and evolution, I will now go forward with my arguments. As I see it, media hybridity is a more fundamental reconfiguration of media universe than multimedia. In both cases we see a "coming together" of multiple media types. However, *multimedia does not threaten the autonomy of different media. They retain their own languages, i.e. ways of organizing media data and accessing and modifying this data*. The typical use of multiple media on the Web or in PowerPoint presentations illustrates this well. Imagine a typical web page from the 2000s that consists of text and video clips embedded somewhere on the page. Both text and video remain separate on every level. Their media languages do not spill into each other. Each media type continues to offer us its own interface. With text, we can scroll up and down and zoom using the browser controls; we can change the browsing settings so it is displayed in a different font. With video, we can use its interface to play it, pause or rewind it, loop a part, and change sound volume. In this example, different media are positioned next to each other but their interfaces and techniques do not interact. This, for me, is a typical example of multimedia.

In contrast, in hybrid media the languages of previously distinct media come together. They exchange properties, create new structures, and interact on the deepest levels. For instance, in *motion graphics* text takes on many properties which were previously

---

[8] Franco Moretti, *Graphs, Maps, Trees: Abstract Models for Literary History* (London Verso, 2007).

unique to cinema, animation, or graphic design. To put this differently, while retaining its old typographic dimensions of font family, size, or line spacing, text also acquires new expressive possibilities from cinema and computer animation. As a word moves closer to us, it can appear out of focus—as though it is a physical object shot by a twentieth-century film camera lens. At the same time, it can now fly in a virtual space, performing physically impossible moves—as any other 3D computer graphics object. Its proportions change depending on what virtual lens the designer has selected. The individual letters that make up a text string can be exploded into many small particles; and so on. In short, in the process of hybridization, the language of typography does not stay as it is. Instead we end up with a new *metalanguage* that combines the techniques of all previously distinct languages, including that of typography.

Another way to distinguish between "multimedia" and "hybrid media" is by noting whether or not the original structure of media data is affected when different media types are combined. For example, when video appears in multimedia documents such as MMS messages, emails sent in HTML format, web pages, or PowerPoint presentations, the structure of video data does not change in any way. Just as with twentieth-century film and video technology, a digital video file is a sequence of individual frames, which have the same size and proportions. Accordingly, the standard methods for interacting with this data type also do not challenge our idea of what "video" is. Like with VCR media players of the 1980–1990s, when the user selects "play," the frames quickly replace each other producing the effect of motion. Video, in short, remains video.

This is typical of multimedia. An example of how some media structure can be reconfigured—the capacity that I take as one of the identifying features of media hybrids—is provided by *The Invisible Shape of Things Past*, a well-known digital cultural heritage project about Berlin's history developed by Joachim Sauter and Dirk Lüsebrink from the media company Art+Com between 1995 and 2007.[9] In this project, film clips become solid objects positioned in a virtual 3D space. Each "film object" is made from individual film frames situated behind one another to form a 3D stack. The angles

[9] http://www.artcom.de/en/projects/project/detail/the-invisible-shape-of-things-past/

between frames and the sizes of individual frames are determined by the parameters of the camera that originally shot the film. We can interact with these new "film objects" as with any other objects in 3D space, flying around using virtual camera controls.

At the same time it is still possible to "see the movie," using the frame stack as a video player. But this operation of access has been rethought. When a user clicks on the front frame in a stack, the subsequent frames positioned behind one another are quickly deleted. You simultaneously see the illusion of movement as in the twentieth century, and the virtual 3D object shrinking at the same time.

In this example of media restructuring, which characterizes media hybridity, the elements that make up the original film's "data structure"—individual frames—have been placed in a new configuration. The old structure has been remapped into a new structure. This new structure retains the original data and their relationship—that is, individual film frames are still organized into a sequence. But it also has new dimensions—size of frames and their angles. The new structure also enables a new type of interface for movie access, which combines virtual space attributes and cinema attributes.

I hope that this discussion makes it clear why hybrid media is not multimedia, and why we need this new term. The term "multimedia" captured the phenomenon the content of different media coming together—but not their languages. Similarly, we cannot use another term that has been frequently used in discussions of computational media—"convergence." The dictionary meanings of "convergence" include "to reach the same point" and "to become gradually less different and eventually the same." But this is not what happens with media languages as they hybridize. Instead, they acquire new properties—becoming richer as a result. For instance, in motion graphics, text acquires the properties of computer animation and cinematography. In 3D computer graphics, rendering of 3D objects can use the techniques of traditional painting. In virtual globes such as Google Earth and Microsoft Virtual Earth, representational possibilities and interfaces for working with maps, satellite imagery, 3D buildings, and photographs are combined to create new richer hybrid representations and new richer interfaces.

In short, "softwarization" of old media did not lead to their "convergence." Instead, after representational formats of older

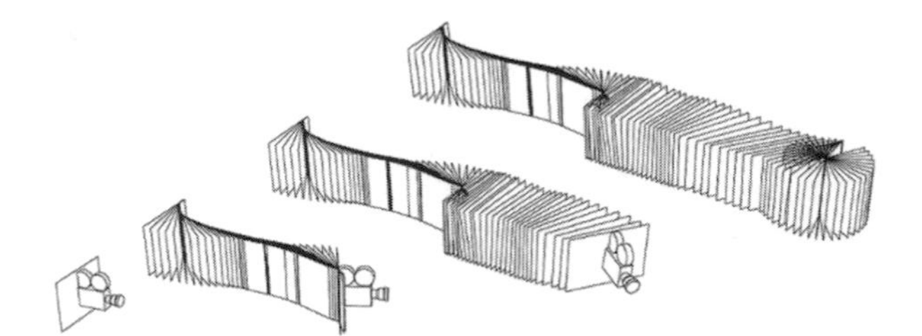

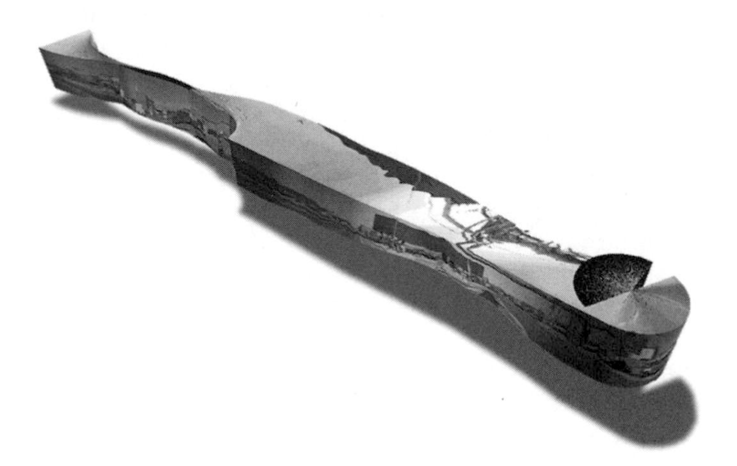

The Invisible Shape of Things Past. *Joachim Sauter and Dirk Lüsebrink (Art+Com), 1995. Bottom: A "film object" consisting of the frames making up a film clip. Top: The angles between frames correspond to the position of the camera. Next three pages: Interaction with a film object in* Invisible Shape *application.*

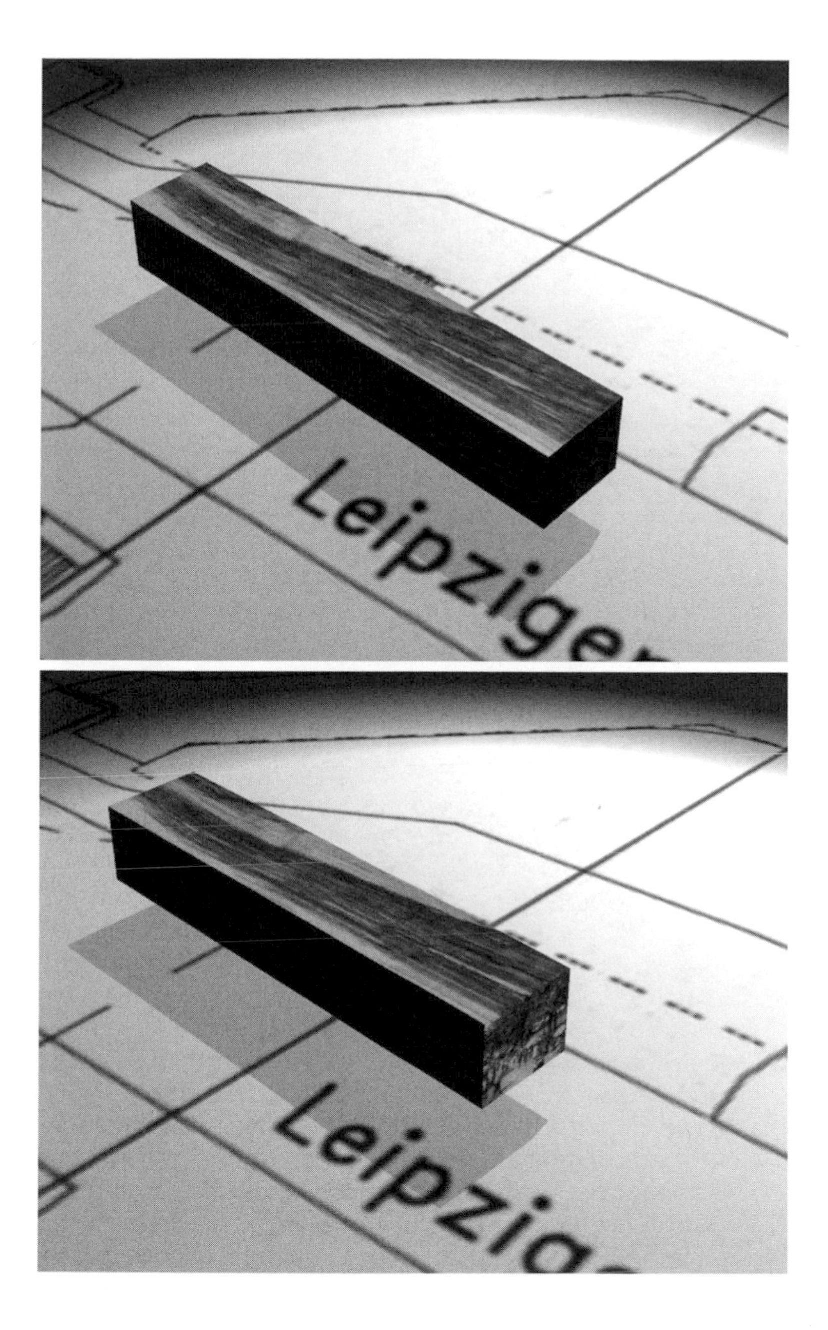
Leipzige
Leipzig

media types, the techniques for creating content in these media and the interfaces for accessing them were unbundled from their physical bases and translated into software, these elements started interacting to produce new hybrids.

This, for me, is the essence of the new stage of computer metamedium development. *The unique properties and techniques of different media have become software elements that can be combined together in previously impossible ways.*

Consequently, if in 1977 Kay and Goldberg speculated that the new computer metamedium would contain "a wide range of already existing and not-yet-invented media," we can now describe one of the key *mechanisms* responsible for the invention of these new media. This mechanism is *hybridization*. The techniques and representational formats of previous physical and electronic media forms, and the new information manipulation techniques and data formats unique to a computer are brought together in new combinations.

In retrospect, it is perhaps not accidental that the publication of Kay and Goldberg's 1977 text, which, for me, summarizes the achievements of the first stage of computational media invention, is directly followed by a seminal project which opens up the next stage—that of hybridization of different media simulated in software. In 1978–9 a group of young researchers at the Architecture Machine Group at MIT (a pre-cursor to the MIT MediaLab) directed by Nicholas Negroponte, constructed Aspen Movie Map—a new type of interactive application that combined a number of media types: video clips, maps, graphics, and diagrams. These different media types were connected through a new type of hypermedia interface. The name of the application—Aspen Movie Map—underscored that this application was neither a map nor a movie but a new hybrid between the two. The project opens the new fundamental stage in the media evolution enabled by its "softwarization"—the stage of hybridity.

The evolution of a computer metamedium

I will continue exploring the metaphor of biological evolution. As the title of Darwin's *On the Origin of Species* (1859) makes clear, the key goal of his evolutionary theory was to explain how different species develop. Darwin proposed that the underlying mechanism was that of natural selection. In the twentieth century, biologists added a number of other explanations (genetic drift, mutation, etc.[10]). While the mechanisms responsible for the development of the computer metamedium are certainly different, we can use the basic ideas of evolutionary theory—emergence of new species over time, with a gradual increase in their number. But even if we only take these basic ideas, there are important differences between biological and media evolution.

For example, if we compare the computer metamedium's development to a biological evolution, we can think of particularly *novel combination of media types* as *new species*. [11] In biological evolution, the emergence of new species is a very slow and gradual process, which requires many generations.[12] Small genetic changes accumulate over long periods before new species emerge. However, new "media species" can emerge overnight—it only requires a novel idea and some programming. Given that today a programmer/designer can use multiple software libraries for media manipulation, and also specialized high-level programming languages specifically designed for rapid testing of ideas and experimentation (Pure Data, Processing, etc.), a talented person can invent new species of media in a few hours.

In evolutionary biology, species are defined as groups of organisms. In media evolution, things work differently. Some novel combinations of media types may appear only once or twice. For instance, a computer science paper may propose a new interface design; a designer may create a unique combination for a particular design project; a film may combine media techniques in a novel

---

[10] http://en.wikipedia.org/wiki/Evolution#Mechanisms (February 6, 2012).

[11] I am aware that not only the details, but also even most of the fundamental assumptions underlying evolution theory continue to be actively debated by scientists. In my references to evolution, I use what I take to be a few commonly accepted ideas from evolutionary theory. While these ideas are being periodically contested and eventually may be disproved, at present they form part of the public "common sense": a set of widely held ideas and concepts about the world.

[12] "Natural selection is the gradual, nonrandom process by which biological traits become either more or less common in a population as a function of differential reproduction of their bearers. It is a key mechanism of evolution." http://en.wikipedia.org/wiki/Natural_selection (February 7, 2012).

way. Imagine that in each case, a new hybrid is never replicated again. This happens quite often.

Thus, some media combinations that emerge in the course of media evolution will not be "selected." Other combinations, on the other hand, may survive and will successfully "replicate." (Again, remember that I am evoking the biological model only as a metaphor, and that no claims are being made that the actual mechanisms of media evolution are similar to the mechanisms of biological evolution.) Eventually such successful hybrids may become the common conventions in media design; built-in features of media authoring applications; commonly used features in social media sites; widely used design patterns; and so on. In other words, they may become new basic building blocks of the computer metamedium that can now be combined with other blocks.

An example of such a successful combination of media "genes" is an "image map" technique for web design. This technique emerged in the middle of the 1990s and was quickly adopted in numerous interactive media projects, games, and websites. How does it work? A photograph, a drawing, a color background, or any other part of a screen is divided into a few invisible parts. When a user clicks inside one of the parts, this activates a hyperlink connected to this part.

As a hybrid, an "image map" combines the technique of hyperlinking with all the techniques for creating and editing still images. Previously, hyperlinks were only attached to a word or a phrase of text and they were usually explicitly marked in some way to make them visible (typically by underlining). When designers started attaching hyperlinks to parts of continuous images or other parts of a web page without explicitly showing them, a new "species" of media was born.

As a new species, it defines new kinds of user behavior and it generates a new experience of media. Rather than being immediately presented with clearly marked, ready to be acted upon hyperlinks, a user now has to explore the screen, mousing over and clicking until s/he comes across a hyperlinked part. Rather than thinking of hyperlinks as discrete locations inside a "dead" screen, a user comes to think of the whole screen as a live interactive surface. On an experiential level, rather than imagining a hyperlink as something which is either present or absent, a user may now experience it as a continuous dimension, with some parts of a surface being "more" strongly hyperlinked than others.

Another example of a successful hybrid that survived and replicated in the course of recent media evolution is a *virtual camera* model used in 3D computer animation. Developed in the 1980s for creating computer animation sequences for feature films, a virtual camera model has gradually become one of the most widely used elements of digital media deployed in video games, virtual environments, program interfaces, feature films, motion graphics, etc.[13]

As we will see in detail in the next part of the book, the new language of visual design (a category which includes graphic design, web design, interface design, motion graphics, and other design areas) that emerged in the second part of the 1990s offers a particularly striking example of media hybridization. Working with software applications, a designer can combine any of the techniques of graphic design, typography, painting, cinematography, animation, computer animation, vector drawing, and 3D modeling. At the same time, s/he also can use many algorithmic techniques for generating new images and forms (such as particle systems or procedural modeling) and transforming them (for instance, by using filters and other digital image processing techniques, which do not have direct equivalents in previous physical, mechanical or electronic media). All these techniques are easily available within a small number of media authoring programs (Photoshop, Illustrator, Flash, Maya, Final Cut, After Effects, various HTML editors, etc.) and they can be easily combined within a single design.

The result is the new design language used today in a large number of countries around the world. The new "global aesthetics" celebrates media hybridity and uses it to engineer emotional reactions, drive narratives, and shape user experiences. The ability to combine previously incompatible techniques of different media is the single common feature of millions of designs being created yearly by professionals and students alike, seen on the web and in print, on big and small screens, in built environments, and all other platforms.

---

[13] For the analysis of a virtual camera use, see Mike Jones, "Vanishing Point: Spatial Composition and the Virtual Camera," *Animation* 3, no. 2 (2007): pp. 225–43.

Like the post-modernism of the 1980s and the web revolution of the 1990s, the "softwarization" of media (the transfer of techniques and interfaces of all previously existing media technologies to software) has flattened history—in this case, the history of modern media. That is, while the historical origins of all the building blocks that make up the computer metamedium—or a particular hybrid—may still be important in some cases, these are now exceptions rather than the rule. Clearly, for a media historian such as myself, the historical origins of all techniques now available in media authoring software do matter. They may also matter for the people encountering a particular media design—but only if a designer chooses to foreground this. For instance, in the logo sequence for DC Comics (Imaginary Forces, 2005) designers used exaggerated artifacts of print and film to evoke particular historical periods in the twentieth century. But when we consider the actual process of media design—the ways in which designers work to go from a sketch or a storyboard or an idea in their head to a finished product—these historical origins no longer matter. When a designer opens his/her computer and starts working, it does not matter whether the technique was originally developed as a part of the simulation of physical or electronic media. Thus, digital paint brushes, filters simulating various natural textures, a camera pan, an aerial perspective, splines, and polygonal meshes, blur and sharpen filters, particle systems, etc.—all have equal status as the building blocks for new hybrids.

Thirty years after Kay and Goldberg predicted that the new computer metamedium would contain "a wide range of already existing and not-yet-invented media," we can see clearly that their prediction was correct. The computer metamedium has indeed been systematically expanding. However, as we now can see, this expansion should not be understood as the simple addition of more and more new "mediums." While a number of new software mediums have been invented in the fifty years following Sketchpad, their number is probably less than one dozen. The key process in the evolution of the computer metamedium involves innovation on a more local level—the previous media types simulated in software (text, sound, drawing, etc.), the techniques for their manipulation, and new computer-native techniques entering in new combinations, creating a much larger number of new "species."

To restate this: following the first stage where most already existing media were simulated in software and a number of new computer techniques for generating and editing of media were invented—the stage that conceptually and practically has been largely completed by the late 1970s—we entered a new period governed by hybridization. The already simulated mediums started exchanging properties and techniques. As a result, the computer metamedium came to contain endless new species. In parallel, we do indeed see a continuous process of the invention of the *new*—but what are being invented are not whole new media types but rather new elements (new techniques for creating, modifying and sharing media data). As soon as they are invented, these new elements start to interact with other, already existing elements. Thus, the processes of invention and hybridization are closely linked and work together.

This, in my view, is the key mechanism responsible for the evolution and expansion of the computer metamedium from the late 1980s until now—and I do not see any reason why this mechanism would become less important in the future. And while at the time when Kay and Goldberg were writing their article the process of hybridization was just barely starting—the first truly significant media hybrid was Aspen Movie Map project created at MIT's Architecture Machine Group in 1978–9—today it is what media design is all about. Thus, from the point of view of today, *the computer metamedium is indeed an umbrella for many things—but rather than only containing a set of separate mediums, it also contains a larger set of smaller building blocks that unite to create hybrids*. These building blocks include algorithms for media creation and editing, data formats, interface metaphors, navigation techniques, physical interaction techniques, web technologies, and other element types. Over time, new elements are being invented and they also become parts of the computer metamedium. Periodically people figure out new ways in which some of the elements available can work together, producing new species. Some of these species may survive. Some may become new conventions, so omnipresent that they are not perceived anymore as combinations of elements which can be taken apart. Still others are forgotten—only to be sometimes reinvented later.

Clearly, all the building blocks that together form the computer metamedium do not have equal importance and equal "linking" possibilities. Some are used more frequently than others, entering in many more combinations. The virtual 3D camera model is currently more widespread than, for example, techniques for rendering realistic looking hair or fur. The 3D camera model is built into every 3D animation application and consequently is used in numerous 3D animations and visual effects sequences; it appears in TV commercials, motion graphics, instructional video, and feature films; it is a part of the user interface in all 3D video games; and is also the interface to popular 3D virtual globes such as Google Earth.[14] In contrast, hair and fur algorithms may not be available in every animation package and their applications are also more limited, since only human and animal characters can have hair or fur.

Some of the new inventions may become so important and influential that it seems no longer appropriate to think of them as just elements. Instead, they may be more appropriately called new *media platforms*—or simply new *mediums*.

Mobile media platforms which emerged in the late 2000s—iOS and Android powering both tablets and mobile phones—are perfect examples here. 3D virtual space, the World Wide Web, and geo media (media which includes GPS coordinates) are other examples of such new media platforms popularized in the 1980s, 1990s, and 2000s, respectively. These media platforms fundamentally reconfigure how all other media are understood and how they can be used. Thus, when we add spatial coordinates to media objects (geo media), place these objects within a single global networked hypertext (the web), or when we start using 3D virtual space as a new platform to design not only buildings and industrial objects but also movies and cartoons, the identity of what we think of as "media" changes in fundamental ways. In fact, we can even say that these changes have been as fundamental as the effects of media "softwarization" in the first place.

But is it true? There is no easy way to resolve this question. Ultimately, it is a matter of perspective. For instance, the simulation of existing media in software and the subsequent period of media hybridization has had a much more substantial impact on

---

[14] For a list of other virtual globe applications and software toolkits, see http://en.wikipedia.org/wiki/Virtual_globe (February 7, 2012).

contemporary visual and spatial aesthetics across all design fields (at least so far) than did the invention of the Web and graphical web browsers.

If we are interested in the histories of visual communication, techniques of representation, and cultural memory, I do think that the universal adoption of *software* throughout global culture industries is at least as importance as the invention of print, photography or cinema. But if we are to focus on the social and political aspects of contemporary media culture and ignore the questions of how media looks and what it can represent—asking instead about who gets to create and distribute media, how people understand themselves and the world through media, and how they create and maintain social relations—we may want to put computer *networks* (be the Web of the 1990s, social media of the 2000s, and whatever new yet-to-be-invented forms will come in the future) in the center.

And yet, it is important to remember that without software, contemporary networks would not exist. Logically and practically, software lies underneath everything that comes later.

For example, if I disconnect my laptop from my wireless network right now, I can still continue using most of the applications—including Word to write this sentence. I can also edit images and video, create a computer animation, design a fully functional website, and compose blog posts. (Of course by the time you are reading this, Microsoft may be offering Word only as an online service, but some other word processors which run locally should be still available…)

But if somebody disables the software running the network, it will go dead.[15] In other words, without the underlying software layers *The Internet Galaxy* (to quote the title of the 2001 book by Manuel Castells[16]) would not be possible. And if software was already responsible for the very first ARPANET (Advanced

[15] Since the late 2000s, there has been a gradual movement towards offering more and more functionality in web applications. However, at least today (2013), unless I am in Singapore or Tallinn which are completely covered with free Wi-Fi courtesy of their governments, I never know if I will find a network connection or not, so I would not want to completely rely on the webware.

[16] Manuel Castells, *The Internet Galaxy: Reflections on the Internet, Business, and Society* (Oxford: Oxford University Press, 2001).

Research Projects Agency Network) computer network which linked two remote machines on October 29, 1969 at the Network Measurement Center at UCLA's School of Engineering and Applied Science and Douglas Engelbart's NLS system at SRI International in Menlo Park, California, its importance and variety only increased as networks develop. Thus, a myriad of software technologies are what allows for media to exist on the web in the first place: images and video in web pages, blogs, Facebook and Twitter, media sharing services such as YouTube and Instagram, aerial photography and 3D buildings in virtual globes, etc. Similarly, the use of 3D virtual space as a platform for media design (which will be discussed in detail in the next part) really means using a number of software algorithms that control the virtual camera, position the objects in space, calculate how they look in perspective, simulate the spatial diffusion of light on the surfaces, and so on.

Hybridity: examples

The examples of media hybrids are all around us: they can be found in user interfaces, web applications, mobile apps, visual design, interactive design, visual effects, locative media, interactive environments, digital art, and other areas of digital culture. Here are a few examples that I have deliberately drawn from different areas. Created in 2004 by Stamen Design (San Francisco), Mappr was one the first popular web mashups.[17] It combined a geographic map and photos from Flickr.[18] Using information entered by Flickr users, the application guessed geographical locations where photos were taken and displayed them on the map. (Today similar map interfaces to photo collections that use GPS data captured with photos are available for iPhoto, Instagram, and other photo services and apps.) Since May 2007, Google Maps has offered Street Views that add panoramic photo-based views of city streets to other media types already used in Google Maps.[19] A

---

[17] http://en.wikipedia.org/wiki/Web_mashup (February 7, 2012).
[18] http://stamen.com/projects/mappr (November 3, 2012).
[19] http://maps.a9.com (January 27, 2006).

hybrid between photography and interfaces for space navigation, Street Views allows users to navigate through a space on a street level by clicking on the arrows superimposed on the panoramic photographs.[20]

Starting in 1991, Japanese media artist Masaki Fujihata created a series of projects called *Field Studies*.[21] These projects place video recordings made in particular places within highly abstracted 3D virtual spaces representing these places. Fujihata started to work on *Field Studies* a decade before the term "locative media" made its appearance. As cameras with built-in GPS did not yet commercially exist at that time, the artist created a special video camera which captured geographical coordinates of each interview location—along with the camera's direction and angle while he was video-taping the interview, as well as his movement though space. The artist used this captured information to create a media interface which combined 3D navigable space and video in a unique way.

For instance, to create the interactive installation *Field-Work@ Alsace* (2002),[22] Fujihata recorded a number of video interviews with the people living in and passing through the area around the border between France and Germany. The project confronts us with a black screen, with a number of three-dimensional white lines showing the artist's movement through the area as he was capturing the interviews. As we navigate around the space, the changing perspective views of these lines suggest the shapes of the Alsace terrain. We also see a number of flat rectangles that are positioned at points where each interview was recorded. Each rectangle is situated at a unique angle that corresponded to the angle of the hand-held video camera during the interview. When you click on a rectangle, the corresponding video plays inside.

In my view, *Alsace* represents a particularly interesting media hybrid. It fuses photography (still images which appear inside rectangles), video documentary (video playing once a user clicks inside a rectangle), the locative media (the movement trajectories recorded by GPS) and 3D virtual space. In addition, *Alsace* uses a

---

[20] http://en.wikipedia.org/wiki/Google_Street_View (July 17, 2008).

[21] www.field-works.net/ (January 27, 2006).

[22] http://www.medienkunstnetz.de/works/field-work/ (February 11, 2012).

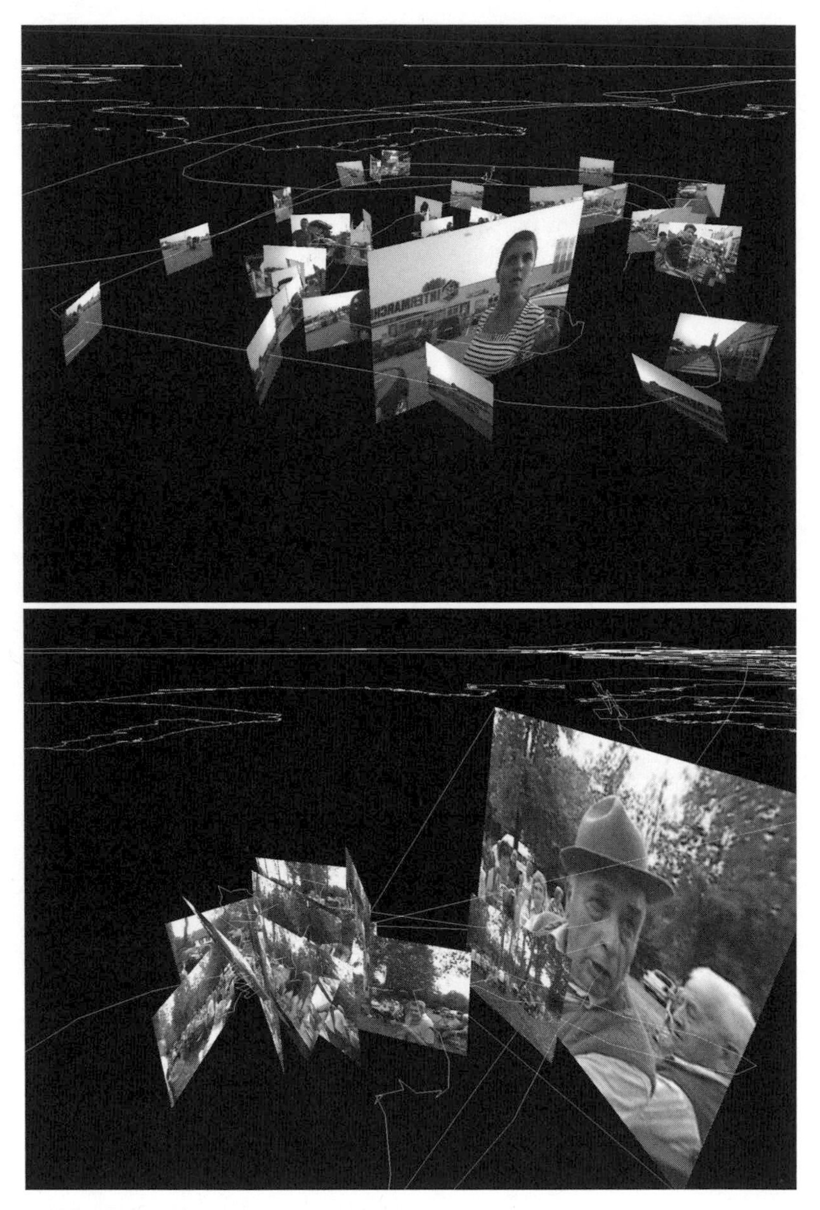

Field-Work@Alsace by *Masaki Fujihata, 2002.*

new media technique developed by Fujihata: the recording not just of the 2D location but also of the 3D orientation of the camera.

The result is a new way to represent collective experiences using 3D space as an overall coordinate system—rather than, for instance, a narrative or a database. At the same time, Fujihata found a simple and elegant way to render the subjective and unique nature of each video interview—situating each rectangle at a particular angle that shows where the camera was during the interview. Additionally, by defining 3D space as an empty void containing only trajectories of Fujihata's movement through the region, the artist introduced the additional dimension of subjectivity. Even today after Google Earth has made 3D navigation of space containing photos and video a common experience, *Alsace* and other projects by Fujihata continue to stand out. They show that to create a new kind of representation it is not enough to simply "add" different media formats and techniques together. Rather, it may be necessary to systematically question the conventions of different media types to make up a hybrid, changing their structure in the process.

A well-known media art project I have already evoked—*The Invisible Shape of Things Past*—also uses 3D space as an umbrella that contains other media types. As I have already discussed, the project maps historical film clips of Berlin recorded throughout the twentieth century into new spatial forms that are integrated into a 3D navigable reconstruction of the city.[23] The forms are constructed by placing subsequent film frames behind each other. In addition to being able to move around the space and play the films, the user can mix and match parts of Berlin by choosing from a number of maps which represent city development in different periods of the twentieth century. Like Alsace, *Invisible Shape* recombines a number of common media types while changing their structure. A video clip becomes a 3D object with a unique shape. Rather than representing a territory as it existed in a particular time, a map can mix parts of the city as they existed at different times.

Another pioneering media hybrid created by Sauter and his company Art+Com is *Interactive Generative Stage* (2002)—a

[23] See www.artcom.de

virtual set whose parameters are interactively controlled by actors during the opera.[24] During the opera performance, the computer reads the body movements and gestures of the actors and uses this information to control the generation of a virtual set projected on a screen behind the stage. The positions of a human body are mapped into various parameters of a virtual architecture such as the layout, texture, color, and light.

Sauter felt that it was important to preserve the constraints of the traditional opera format—actors foregrounded by lighting with the set behind them—while carefully adding new dimensions to it.[25] Therefore, following the conventions of traditional opera the virtual set appears as a backdrop behind the actors—except now it not a static picture but a dynamic architectural construction that changes throughout the opera. As a result, the identity of a theatrical space changes from that of a backdrop to a main actor—and a very versatile actor at that—since throughout the opera it adopts different personalities and continues to surprise the audience with new behaviors. This kind of fundamental redefinition of an element making a new hybrid is rare, but when a designer is able to achieve this, the result is very powerful.

Not every hybrid is necessarily elegant, convincing, or forward-looking. Some of the interfaces of popular media creation and access applications look like the work of an aspiring DJ, mixing operations from old interfaces of various media with new GUI principles in sometimes erratic and unpredictable ways. In my view, a striking example of such a problematic hybrid is the interface of Adobe Acrobat version 8.0, released in November 2006.[26] (Note that since the interfaces of all commercial software applications typically change from version to version, just as elsewhere in the book, this example refers to this particular version of Adobe Acrobat.) This version of Acrobat's User Interface combines interface metaphors from a variety of media traditions and technologies in a way that, at least to me, does not always seem to be logical. Within a single interface, we get 1) the interface elements from analog media

---

[24] The full name of the project is *Interactive generative stage and dynamic costume for André Werner's opera, 'Marlowe, the Jew of Malta.'*

[25] Joachim Sauter, personal communication, Berlin, July 2002.

[26] http://en.wikipedia.org/wiki/Adobe_Acrobat#Version_8.0 (February 8, 2012).

recorders/players of the twentieth century, e.g., VCR-style arrow buttons; 2) interface elements from image editing software, e.g., a zoom tool; 3) interface elements which have strong association with the print tradition—although they never existed in print (page icons also controlling the zoom factor); (4) elements which have existed in books (the bookmarks window); (5) the standard elements of a GUI such as search, filter, and multiple windows. It seems that Acrobat designers wanted to give users a variety of ways to navigate through documents. However, I find the use of so many navigation metaphors confusing. For instance, given that Acrobat was designed to closely simulate the experience of print documents, it is not clear to me why I am asked to move through the pages by clicking on the forward and backward arrows—an interface convention which is normally used for moving image media.

The hybrids do not necessarily have to involve a "deep" reconfiguration of previously separate media languages and/or the common structures of media objects—the way, for example, *The Invisible Shape* reconfigures the structure of a film object. Consider web mashups which "combine data elements from multiple sources, hiding this behind a simple unified graphical interface.[27] I have already used one example of a Web mashup—Mappr. Here are some other successful early examples. *Flickrvision 3D* (David Troy, 2007) used data provided by Flickr and the virtual globe from Poly 9 FreeEarth to create a mashup which continually showed the new photos uploaded to Flickr attached to the virtual globe at those locations where the photos were taken. Another mashup called *Liveplasma* (2005) used Amazon services and data to offer a music and discovery engine. When a user selected an actor, a movie director, a movie title, or a band name, *Liveplasma* generated an interactive map that showed related actors, movie directors, etc. using various dimensions such as style, influences, popularity, and others.[28] Although *Liveplasma* suggests that the purpose of these maps is to lead you to discover the items that you are also likely to like (so you purchase them on amazon.com), these maps are valuable in themselves. They employ newly available rich data about people's cultural preferences and behavior collected by Web 2.0 sites

[27] http://en.wikipedia.org/wiki/Mashup_(web_application_hybrid) (July 19, 2008).
[28] http://www.liveplasma.com/ (August 16, 2008).

such as Amazon to do something that was not possible until the 2000s. That is, rather than mapping cultural relationships based on the ideas of a single person or a group of experts, they reveal how these relationships are understood by actual cultural consumers.

The development of mashups is supported by the gradually growing number of web APIs offered by a variety of companies. An API provides an easy way for a programmer to create new programs, which use services or data provided by web companies. For example, you can use the Google Maps API to generate interactive Google maps inside your website. When a user comes to the site and enters an address into the map interface, Google servers get the request and send back the new map. When I checked the mashup tracker programmableweb.com on February 8, 2012, it listed 2,337 mashups that use the Google Maps API.[29] Many mashups combine between half a dozen and a dozen APIs from different services. By September 2012, the site tracked over 7,000 different APIs.[30] Sorted by the number of mashups using these APIs, the top entries were the Google Maps API used in 2,416 mashups, the Twitter API used in 717 mashups, and YouTube, used in 650 mashups.[31] These numbers may only represent a small percentage of all mashups out there—just think about all the times you encounter a Google map used as part of some website or service. According to one 2012 estimate, the Google Maps API was used in 350,000 web sites. Thus, the numbers reported by programmableweb.com perhaps only indicate the relative proportions in APIs use in mashups.

Visually, many mashups may appear as typical multimedia web pages—but they are more than that. As the Wikipedia article on "mashup (web application hybrid)" explains, "A site that allows a user to embed a YouTube video for instance, is not a mashup site... the site should itself access 3rd party data using an API, and *process that data in some way to increase its value* to the site's users." (My emphasis). Although the terms used by the authors—processing data to increase its value—may appear to be strictly business-like, they also capture the difference between multimedia and hybrid media

[29] http://www.programmableweb.com/apis/directory/1?sort=mashups (February 8, 2012).

[30] http://blog.programmableweb.com/2012/08/23/7000-apis-twice-as-many-as-this-time-last-year/ (August 23, 2012).

[31] http://www.programmableweb.com/apis/directory/1?sort=mashups (November 4, 2012).

in a theoretically accurate way. Paraphrasing the article's authors, we can say that in the case of successful artistic hybrids such as *The Invisible Shape* or *Alsace*, separate representational formats (video, photography, 2D map, 3D virtual globe) and media navigation techniques (playing a video, zooming into a 2D document, moving around a space using a virtual camera) are brought together in ways which *increase the representational and expressive valu*e offered by each media type used. However, in contrast to the web mashups that started to appear *en masse* in 2006 when Amazon, Flickr, Google and other major web companies offered public API (i.e., they made it possible for others to use their services and some of their data—for instance, using Flickr images as a part of a mashup), these projects also used their own data, which the artists carefully selected or created themselves. As a result, the artists have much more control over the aesthetic experience and the "personality" projected by their works than an author of a mashup, which relies on both data and the interfaces provided by other companies and non-profit organizations such as OpenLayers.

I am not trying to criticize the web mashup technology—I only want to suggest that if the project's goal is to put forward a different representational model and a unique aesthetic experience, choosing from the same set of web sources and data sources available to everybody else may be not the right solution. And the argument that web mashup author acts as a DJ who creates by mixing what already exists also does not work here—since a DJ has both more control over the parameters of the mix, and more recordings to choose from.

In discussing examples of hybrids so far I have implicitly presented them as combinations and reconfigurations of previously existing media, which include both the simulations of physical media such as print and new media such as computer animation. In other words, I have relied on the notion consistent with Kay's own formulation that the computer mediamedium can be understood as a collection of different mediums. For example, I talked about how *Alsace* combined photography and video documentary (pre-digital media simulated in a computer) with locative data and 3D virtual space (new computer media).

However, we can also think of media hybridity using a different conceptualization of the metamedium. That is, rather than emphasizing "whole" media we can focus on their building blocks—i.e., different types of media data (or "media content") and two types of techniques which can operate on this data. (I called these techniques "media-specific," if they can only work on specific media types; I called "media-independent" those techniques that are implemented to work with many types).

From this perspective, the new *hybrid media species*—a single project, web service, or a software program—represents the meeting of various techniques that previously belonged to different mediums. Chapter 5 will develop this concept in detail using examples from motion graphics and design. We will also see how hybridization as enabled by software became one of the dominant aesthetics of contemporary media. But as a way of getting started, let us take the example of hybridization and discuss it in terms of data types and data manipulation techniques. For this example, I will use an application which should be familiar to everybody and which I have already briefly evoked a number of times—Google Earth.

Google Earth is based on an earlier application called Earth Viewer developed by Keyhole, Inc. This company was acquired by Google in 2004. In its turn, Earth Viewer took the idea of seamless interactive navigation cinema-style around the detailed and hybrid spatial representation from 1996 project *Terravision*. The project was created by the same innovative design team which is also responsible for some of the other outstanding media hybrids we already encountered—Joachim Sauter and Art+Com.[32] (The following analysis applies the features and interfaces in Google Earth 5, released in May 2009.) Using this application, you can navigate around the Earth's surface, zooming in and out; turn on and off a variety of data overlays; search for places and directions; mark places on the map and share these additions with all other users of Google Earth; import your own information including images and GPS data; create movies of touring around; and more.

When Google Earth was first released in June 2005, Google called it a "3D interface to the planet."[33] This description itself tells us right away that we are not dealing with a twentieth-century map or any other representation already familiar to users. So what are

[32] http://www.artcom.de

[33] http://windowssecrets.com/langalist-plus/a-3d-interface-to-the-planet/ (February 10, 2012).

the key elements of the experience offered by this "3D interface to the planet" that make it stand out from the variety of other cultural applications that allow a user to navigate around and perform actions on some data—2D Google Maps, web browsers, iTunes, educational multimedia applications, and so on? These elements are both its hybrid terrain and the corresponding hybrid navigation mechanisms.

The representation of Earth's surface that appears in the main Google Map window, called "3D Viewer," combines satellite photography, 3D elevation data, 3D models of buildings, and the graphics elements familiar to us from modern paper maps (vector graphics and text labels identifying roads, country boundaries, etc.). Importantly, the four types of data are "glued together," (i.e., rendered directly on top of each), thus appearing as a single visual source. This is a perfect example of a hybrid. The different media types are brought together to create a new representation.

The 3D interface offered by Google Earth is also a hybrid. It draws on the new type of computational media that has been evolving since the late 1960s: 3D interactive navigable space. It also uses the computer simulation of Hollywood cinematography techniques developed in the 3D computer animation field since the 1970s. The user navigates around the hybrid terrain using a set of defined camera controls that extend the language of zoom, tilt, and pan, developed in film cinematography. (Google Earth 6 defines the following "3D navigation techniques": move left, right, up and down, rotate clockwise and counter wise, tilt up and down, zoom in and out, zoom + automatic tilt, look and reset.[34])

In addition to this basic navigation system, the application also offers a more explicitly cinematic and more automatic method called Touring, where the camera flies between the points in a seamless trajectory.

(I am distinguishing between "3D interactive space" and "simulated camera" for the following reason. While software used by computer animators, game designers, and media designers provides a virtual camera interface to 3D space with all traditional cinematographic controls, other applications which also use virtual spaces such as VR and computer games do not. Instead they use

[34] http://support.google.com/earth/bin/answer.py?hl=en&answer=148115&topic=2376154&ctx=topic (February 10, 2012).

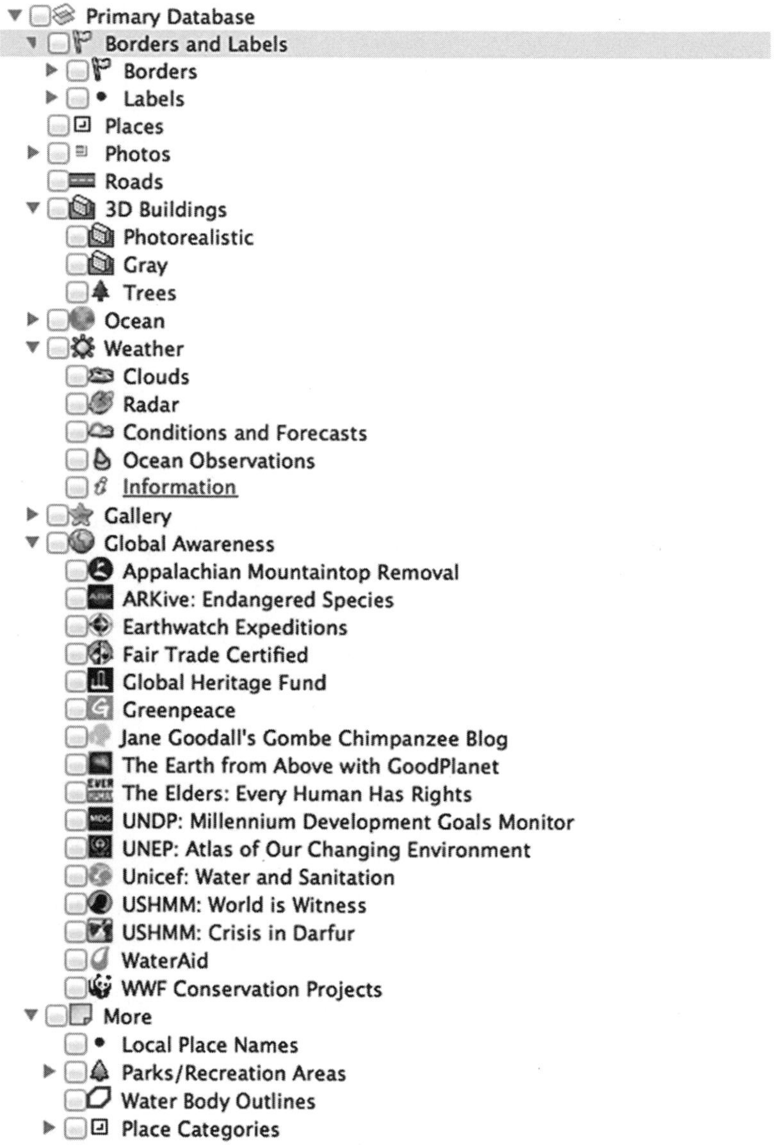

- Primary Database
  - Borders and Labels
    - Borders
    - Labels
  - Places
  - Photos
  - Roads
  - 3D Buildings
    - Photorealistic
    - Gray
    - Trees
  - Ocean
  - Weather
    - Clouds
    - Radar
    - Conditions and Forecasts
    - Ocean Observations
    - Information
  - Gallery
  - Global Awareness
    - Appalachian Mountaintop Removal
    - ARKive: Endangered Species
    - Earthwatch Expeditions
    - Fair Trade Certified
    - Global Heritage Fund
    - Greenpeace
    - Jane Goodall's Gombe Chimpanzee Blog
    - The Earth from Above with GoodPlanet
    - The Elders: Every Human Has Rights
    - UNDP: Millennium Development Goals Monitor
    - UNEP: Atlas of Our Changing Environment
    - Unicef: Water and Sanitation
    - USHMM: World is Witness
    - USHMM: Crisis in Darfur
    - WaterAid
    - WWF Conservation Projects
  - More
    - Local Place Names
    - Parks/Recreation Areas
    - Water Body Outlines
    - Place Categories

The Layers window in Google Earth 7.0 (2012). The screenshot shows only some of the hundreds of layers available.

movements of the body, head, or fingers of the human user to guide a camera. Thus, a 3D space representation and a 3D camera model do not have to go together.)

While Google Earth's core data model (satellite imagery + elevation data + map symbols) remains unchanged over time and a number of software releases, the continuing addition of new data sources and data types makes the representation increasingly rich—and simultaneously increases its hybridity. These additional types of data include links to web content, Street View (launched on May 25, 2007), normal and panoramic hi-res photos, historical imagery (in Google Earth 5.0), underwater terrain, the Moon, Mars, real-time traffic, etc. The addition of some of these new data types requires parallel addition of new navigation mechanisms. Thus, side-by-side with the original core 3D cinema-like interface we now find other interfaces.

In this way, the techniques for working with data provided by Google Earth also become more hybrid. When interacting with a 3D building, a user can "swoop to the top or side" of the building. In the case of high-res gigapixel photos, Google Earth offers a special way of "flying into" a photo which then can be panned and zoomed. And with Street View, yet another set of navigation techniques is provided.[35]

Strategies of hybridization

As we see, media hybrids can be structured in different ways. In user interfaces such as the interface of Acrobat Reader, the operations which previously belonged to specific physical, mechanical, and electronic media are combined to offer the user more ways to navigate and work with the computer documents (combination of different *interface techniques*). Google Earth combines different types of media to provide more comprehensive information about places when either media can do by itself (a combination of *media*

[35] "Using the keyboard or mouse the horizontal and vertical viewing direction and the zoom level can be selected. A solid or broken line in the photo shows the approximate path followed by the camera car, and arrows link to the next photo in each direction. At junctions and crossings of camera car routes, more arrows are shown." http://en.wikipedia.org/wiki/Google_Street_View (February 11, 2012).

types). Mappr exemplifies another strategy: using a 2D geo map as an interface to a media collection—in this case, photos uploaded on Flickr (using *one media type as an interface to another media type*.) *Alsace* and *Invisible Shape* exemplify yet another type of media combination: using *one media type as an enclosure for another media type* (3D virtual space containing film and video and clips).

A complementary way of categorizing media hybrids is by asking if a particular hybrid offers new ways of representing the world, and/or new ways of navigating these representations. Hybrids may combine and/or reconfigure familiar media formats and media interfaces to offer *new types of hybrid representations*. For instance, Google Earth and Microsoft Bing Maps combine different media types and interface techniques to provide more comprehensive information about places than either media can do by itself. The ambitions behind *Alsace* and *Invisible Shape* are different—not to provide more information by combining existing media formats but rather to reconfigure these formats in order to create new representations of human collective and individual experiences that fuse objective and subjective dimensions. But in both cases, we can say that the overall goal is to *represent the world or our experience in a new way by combining and possibly reconfiguring* already familiar media representations (photos, video, maps, 3D objects, web pages, panoramic photos, etc.) Another good example of such hybrids is Microsoft Photosynth which offers new types of 3D representations ("synths") made by matching many photographs of the same scenes—such as a detailed model of a 3D Notre Dame Cathedral created entirely from its photos on Flickr.[36]

Secondly, the hybrids may focus on *new ways of navigation and interaction with already existing media formats*. Here the media type itself is neither modified nor combined with other media—instead, hybridization happens in the UI and the tools provided by the project, service or the application for working with this media type. For example, in the case of Mappr, both 2D map and photo formats already existed separately. The mashup links them together, turning the map into an interface to the photos available on Flickr.[37]

---

[36] http://www.ted.com/talks/blaise_aguera_y_arcas_demos_photosynth.html (February 19, 2012).

[37] This mashup also exemplifies an important development within metamedia

(Flickr itself later offered the similar map interface[38] as well as an application to allow users to place their photos on a world map.[39] On February 19, 2012, Flickr's map interface contained over 175 million geo-tagged photos.)

In summary, a hybrid may define new navigation and interaction techniques that operate over non-modified media formats. Alternatively, a hybrid may define new media formats but use already existing interaction/interface techniques. A *hybrid may also combine both strategies, i.e. it can define both new interfaces/tools and new media formats at the same time.* This, however, requires a real creativity and deep understanding of both media computing and media aesthetics, so such hybrids do not appear very often. (*Alsace*, *Invisible Shape,* and Photosynth are able to combine both strategies, and this is why for me they stand out from the multitude of new media projects and applications created in the last two decades.)

You may notice that the distinction between a "representation" (or a "media format") and an "interface/tool" corresponds to the two fundamental components of all modern software: *data structures* and *algorithms*. This is not accidental. Each tool offered by a media authoring, editing or viewing application corresponds to an algorithm that either processes the data in a particular format in some way, or generates new data in this format. For example, let us assume that our media format is a photo (or, more generally, a raster image). To generate a gallery view of the photos an algorithm has to process each photo to fit it into a specified size (this is done by calculating averages of groups of pixels and using a new smaller set of these average values). To draw a line over a photo requires calling another algorithm that calculates new colors for the pixels beneath the line. Thus, "working with media" using application software essentially means running different algorithms over the data.

---

evolution: a convergence between media and spatial data. The three main forms of this convergence are: 1) a 2D map used as an interface to other media types (as in Mappr); 2) a 3D virtual map used as an interface to other media types (as in *Alsace, Invisible Shapes* or Google Earth); 3) location information automatically added by a capture device to images and video recordings.

[38] http://www.flickr.com/map/

[39] http://en.wikipedia.org/wiki/Flickr#Organizr (March 2, 2012).

While this logical differentiation is clear and useful for the person who understands programming, when we consider the user's experience of media authoring/viewing/cataloging/sharing applications, web services, and interactive media projects, it is harder to maintain. In the world of application software, media data and interfaces/tools never exist in isolation from each other. Unless you know how to program, you never encounter media content types—digital photos, digital videos, maps, etc.—by themselves. Instead, you encounter media content through particular software applications, or the custom interfaces defined by the designers of a particular project. In other words, you always work with data in a context of some application, one that comes with its own interface and tools. This means that *as experienced by a user of application software, "representation" consists of two interlinked parts: data structured in particular ways and the interfaces/tools provided to navigate and work with this data*. (The same applies for the concept of "information".) For example, a "3D virtual space" as it is defined in 3D computer animation and CAD applications, computer games, virtual globes, and other applications is not only a set of coordinates that make up 3D objects and a perspective transformation but also a set of navigation methods—i.e., a virtual camera model. A "photograph" as defined by media editing applications includes various editing operations that can be performed on it such as scale, cut and paste, make mask, add layers, etc. Liveplasma's interactive culture maps are not only relationships between the items on the map that we can see but also the tools provided to construct and navigate these maps. And the unique "Earth" in Google Earth is made up not only from its hybrid data model (satellite photography, elevation, 2D map, 3D buildings, panoramas) but also the rich techniques for navigating and exploring this data.

CHAPTER FOUR

Soft evolution

Algorithms and data structures

What makes possible the hybridization of media creation, editing, and navigation techniques? To start answering this question we need to ask once again what it means to simulate physical media in software. For example, what does it mean to simulate photography or print media?

A naïve answer is that computers simulate the actual media objects themselves. For example, a digital photograph simulates an analog photograph printed on paper; a digital illustration simulates an illustration drawn on paper; and digital video simulates analog video recorded on videotape. But that is not how things actually work.

What software simulates are the physical, mechanical, or electronic *techniques used to navigate, create, edit, and interact with media data.* (And, of course, software also extends and augments them, as discussed in detail in Part 1.) For example, the simulation of print includes the techniques for writing and editing text (copy, cut, paste, insert); the techniques for modifying the appearance of this text (change fonts or text color) and the layout of the document (define margins, insert page numbers, etc.); and the techniques for viewing the final document (go to the next page, view multiple pages, zoom, make bookmark). Similarly, software simulation of cinema includes all the techniques of cinematography such as user-defined focus, the grammar of camera movements (pan, dolly, zoom), the particular lens that defines what part of a virtual scene the camera will see, etc. The simulation of analog video includes a set of navigation commands: play forward, play in reverse, fast forward, loop, etc. In short: *to simulate a medium in software means to simulate its tools and interfaces, rather than its "material."*

Before their softwarization, the techniques available in a particular medium were part of its "hardware." This hardware included instruments for inscribing information on some material, modifying this information, and—if the information was not directly accessible to human senses such as in the case of sound recording—presenting it. Together the material and the instruments determined what a given medium could do.

For example, the techniques available for writing were determined by the properties of paper and writing instruments, such as a fountain pen or a typewriter. (The paper allows making marks on top of other marks, the marks can be erased if one uses pencil but not pen, etc.) The techniques of filmmaking were similarly determined by the properties of film stock and the recording instrument (i.e. a film camera). Because each medium used its own distinct materials and physical, mechanical, or electronic instruments, each also developed its own set of techniques, with little overlap.

Thus, because media techniques were part of specific incompatible hardware, their hybridization was prevented. For instance, you could white out a word while typing on a typewriter and type over—but you could not do this with the already exposed film. Or, you could zoom out while filming progressively revealing more information—but you could not do the same while reading a book (i.e. you could not instantly reformat the book to see a whole chapter at once.) A printed book interface only allowed you to access information at a constant level of detail—whatever would fit a two-page spread.[1]

Software simulation liberates media creation and interaction techniques from their respective hardware. The techniques are translated into software, i.e. each becomes a separate algorithm.

[1] This was one of the conventions of books which early twentieth-century book experiments by modernist poets, and designers such as Marinetti, Rozanova, Kruchenykh, Lissitzky, and others worked against.

And what about the physical materials of different mediums? It may seem that in the process of simulation they are eliminated. Instead, media algorithms, like all software, work on a single material—digital data, i.e. numbers.

However, the reality is more complex and more interesting. The differences between materials of distinct media do not simply disappear into thin air. Instead of a variety of physical materials computational mediums use different ways of coding and storing information—different data structures. And here comes the crucial point. *In place of a large number of physical materials, software simulations use a smaller number of data structures.*

(A note on my use of the term "data structure." In computer science, a data structure is defined as "a particular way of storing and organizing data in a computer so that it can be used efficiently."[2] The examples of data structures are arrays, lists, and trees. I am going to appropriate this term and use it somewhat differently, to refer to "higher-level" representations which are central to contemporary computational media: a bitmap image, a vector image, a polygonal 3D model, NURBS models, a text file, HTML, XML and a few others. Although the IT, media, and culture industries revolve around these formats, they do not have a standard name of their own. For me the term "representation" is too culturally loaded while the term "data type" sounds strictly technical. I prefer "data structure" because it simultaneously has a specific meaning in computer science and also a meaning in humanities—i.e. the "structure" part. The term will keep reminding us that what we experience as "media," "content" or "cultural artifact" is technically a set of data organized in a particular way.)

Consider all different types of materials that can be used to create 2D images, from stone, parchment and canvas to all the dozens types of paper, that one can find today in an art supply store. Add to those all of the different kinds of photographic film, X-Ray film, film stocks, celluloid used for animation, etc. Digital imaging substitutes all these different materials by employing just two data structures. The first is the bitmapped image—a grid of discrete "picture elements" (i.e., pixels) each having its own color

[2] Paul E. Black, ed., entry for "data structure" in *Dictionary of Algorithms and Data Structures*, U.S. National Institute of Standards and Technology, http://xlinux.nist.gov/dads/

or gray-scale value. The second is the vector image, consisting of lines and shapes defined by mathematical equations.

So what then happens to all the different effects that these physical materials were making possible? Drawing on rough paper produces different effects from drawing on smooth paper; carving an image on wood is different from etching the same drawing in metal. With softwarization, all these effects are moved from "hardware" (physical materials and tools) into software.

All algorithms for creating and editing continuous-tone images work on the same data structure—a grid of pixels. And while they use different computational steps, the end result of these computations is always the same—a modification in the colors of some of the pixels. Depending on which pixels are being modified and in what fashion, the algorithms can visually simulate the effects of drawing on smooth or rough paper, using oils on canvas, carving on wood, and making paintings and drawings using a variety of physical instruments and materials.

If particular medium effects were previously the result of the interaction between the properties of the tools and the properties of the material, now they are the result of different algorithms modifying a single data structure. So we can first apply the algorithm that acts as a brush on canvas, then an algorithm that creates an effect of a watercolor brush on rough paper, then a fine pen on a smooth paper, and so on. In short, the techniques of previously separate mediums can now be easily combined within a single image. And since media applications such as Photoshop offer dozens of these algorithms (presented to the user as tools and filters with controls and options), this theoretical possibility becomes a standard practice. The result is a new hybrid medium that combines the possibilities of many once-separate mediums.

Instead of numerous separate materials and instruments, we can now use a single software application whose tools and filters can simulate different media creation and modification techniques. The effects that previously could not be combined since they were tied to unique materials are now available from a single pull-down menu. And when somebody invents a new algorithm, or a new version of already existing algorithm, it can easily be added to this menu using the plug-in architecture that became standardized in the 1990s (the term "plug-in" was coined in 1987 by the developers of

Digital Darkroom, a photo editing application[3]). And, of course, numerous other image creation and modification techniques that did not exist previously can be also added: image arithmetic, algorithmic texture generation (such as Photoshop's Render Clouds filter), a variety of blur filters, and so on (Photoshop's menus provide many more examples).

To summarize this analysis: software simulation substitutes a variety of distinct materials and the tools used to inscribe information (i.e., make marks) on these materials with a new hybrid medium defined by a common data structure. Because of this common structure, multiple techniques that were previously unique to different media can now be used together. At the same time, new previously non-existent techniques can be added as well, so long as they can operate on the same data structure.

(Note: many standard contemporary image formats including the Photoshop .psd format are much more complex than a simple pixel grid—they can include alpha channels, multiple layers, and color profiles; they can also combine bitmapped and vector representations. However, in this discussion, I am only talking about their common denominator—what the algorithms work on when an image is loaded in memory—an array of pixels holding color values.)

Let us now look at another example of what happens with physical materials of different mediums when they are simulated in software. Consider 3D modeling software such as Blender, Maya, 3ds Max Studio, LightWave 3D, or Google's SketchUp. These applications provide the techniques for defining 3D forms which were previously "hardwired" to different physical media. For example, you can use sculpting tools to create a rounded form as though you are using clay. 3D applications also provide dozens of new techniques for defining and modifying forms not available previously: bevel, extrude, spherize, randomize, boolean operations, smooth, loft, morph, simplify, subdivide, and so on.[4] As with image editing software, new techniques can always be added as long as they operate on the standard data structures already used by 3D software. (The most common ones are polygonal models

[3] http://en.wikipedia.org/wiki/Digital_Darkroom (February 19, 2012)

[4] A list of a subset of the operations that work on 3D models made from polygons is provided in http://en.wikipedia.org/wiki/Polygon_modeling#Operations

and NURBS models.[5] The former consist of flat polygons; the latter are defined by smooth curves.) These data structures are the new "materials" that software substitutes for a variety of physical materials used by humans to create physical 3D forms such as stone, wood, clay, or concrete.

These two examples of raster image and 3D model data structures should make clear why it is incorrect to think that computers always work on a single digital material, so to speak, i.e. the binary code made from 0 and 1. Of course this is what happens on a low-level machine level—but this is largely irrelevant as far as application software users and people who write this software are concerned. Contemporary media software contains its own "materials": data structures used to represent still and moving images, 3D forms, volumes and spaces, texts, sound compositions, print designs, web pages, and other "cultural data." These data structures do not correspond to physical materials in a 1:1 fashion. Instead, a number of physical materials are mapped onto a single structure—for instance, different imaging materials such as paper, canvas, photographic film, and videotape become a single data structure (i.e., a bitmapped image). This *many to one mapping from physical materials to data structures is one of the conditions which enables hybridization of media techniques.*

What is a "medium"?

I have spent considerable time analyzing the specificity of media software in relation to pre-digital media. This analysis allows us to better understand why hybridity became the next stage in computer metamedium evolution and to begin tracing the mechanisms of hybridization. Let us now see if we can use our new insights to answer one of the key questions of this book: what is "media" after its softwarization?

To avoid confusion: I am not talking about the actual *content* of media programs and sources, be they television programs, newspapers, blogs, or the terrain in Google Earth. There are already a number of academic disciplines which study

[5] http://en.wikipedia.org/wiki/3D_modeling

media content and its reception: Media Studies, Communication, Journalism, Film and TV Studies, Game Studies, Cultural Studies, and Internet Studies. I am also not going to talk about *media industries*—production, distribution, reception, markets, economic aspects, etc., because academic disciplines already analyze these extensively. However, they usually do not so closely analyze the tools, technologies and workflows used to produce media content. Even when they do this analysis, it is only done in relation to the tools of a particular media field. This is because the modern academic study of culture follows the commercial culture industries' strict division by type of content. Thus, Game Studies looks at games, Film and TV Studies looks at films and television programs, Design Studies looks at design, Internet Studies looks at the web, etc. Because of these divisions, these disciplines ignore the common features of all media and cultural production being done today which are the result of their reliance on the same technology—*application software for media authoring and editing*. (This is one of the reasons why we need a Software Studies perspective—to focus our attention on common cultural patterns related to the use of software technology in all of these diverse cultural fields and media industries.)

I will also bracket media reception—in other words, I will try to define what "media" is today for its creators as opposed to media consumers. (While in the 2000s there were many discussions about the blurring of these definitions because of the falling prices of tools and emergence of social media, in practice they did not get erased.) And finally, people who know computer programming and can create media by writing programs will also have a different understanding of media—but the majority of content creators use only application software.

The universe of the users of application software includes "creative professionals": motion graphics artists, graphic designers, photographers, video editors, product designers, architects, visual artists, etc. It also includes "prosumers" (or "pro-ams") making anime remixes, editing documentary videos which they will upload to YouTube or Vimeo, shooting photos which they will post to their Pro Flickr accounts, or uploading their art images to deviantArt.

I want to understand what it means to create "media" for all these people as defined by the possibilities of the software they are using—Photoshop, Gimp, Illustrator, InDesign, After Effects,

Final Cut, Premiere, CinePaint, Maya, Dreamweaver, WordPress, Blogger, Flash, OpenOffice, Pages, Microsoft Word, Flame, Maya, and so on. (And, speaking of Word and other text processing applications, I should also add millions of people who use it daily and who, therefore, can be considered experts or at least prosumers in at least one medium—that of text authoring and editing.)

Recall one of the dictionary definitions of a "medium" which opens *Understanding Metamedia* (Chapter 2): "A specific kind of artistic technique or means of expression as determined by the materials used or the creative methods involved." (For example, "the medium of lithography.") Thus, different mediums have different techniques, means of expression and creative methods. These differences certainly do not disappear when we switch to software applications. For example, besides the obvious representational and expressive difference between 3D models and moving images, a designer who is modeling a game character in Maya and a designer who is making animation in After Effects will have access to different sets of tools. But are there some conceptual similarities between the way these two designers will be working because they both use media software?

In short: *what is "media" today as defined by software applications used to create and edit it?*

As I have already discussed, earlier physical, mechanical, and electronic media consisted of two components: materials used to hold information and the tools or equipment used to record, edit, and view this information. For example, the "film medium" used film stock for information storage, a film camera for recording, a projector for showing films, and editing devices such as Moviola and Steenbeck. The medium of hand engraving used metal plates (typically copper) to hold information, and special, hardened steel tools to create it by making grooves in the plate.

Do these two components find their analogs in software? Here is one answer which we can give to this question: *Materials become data structures; the physical, mechanical, and electronic tools are transformed into software tools which operate on these data structures.* From this perspective, regardless of the particular media field, all designers and artists working with media software are doing the same thing: using the tools provided by the software to create, modify, and edit data organized in particular data structures.

This answer is compelling but not precise. As I already discussed, many materials are mapped into a single data structure. Thus, the move from physical media to software apps involves a redistribution of the roles previously played by the physical tools and materials. When I use a watercolor brush and a rough-textured paper, the resulting brushstrokes are equally the result of the brush, the liquid, and the paper. But when I use a "watercolor" brush in Photoshop, or apply a "watercolor" filter to an already existing image, the result is determined solely by an algorithm, that modifies the colors of the pixels in a particular manner. The pixels are only memory locations, which hold the color values—they do not have any properties of their own, unlike physical materials.

Therefore, we do not have a one-to-one mapping between physical materials and data structures. The same data structure (such as a bitmap image) can be used to simulate many imaging techniques: from watercolor and engraving to photography. To make this concrete, look at all JPEG images on your computer. Some are your photographs uploaded from a mobile phone or digital camera; others are small graphical icons used by various applications; still others may be the notes you made with a note-taking app, and so on. The same data structures hold multiple media.

This is why rather than stating that materials turn into data structures while tools turn into algorithms, it would be more correct to say that *a medium as simulated in software is a combination of a data structure and set of algorithms*. The same data structure can be shared across multiple medium simulations, but at least some of the algorithms will be unique to each medium.

We have arrived at a definition of a software "medium," which can be written in this way:

Medium = algorithms + a data structure

Algorithms and data structures happen to be two fundamental elements of computer programming. In fact, one of the most influential books in the history of computer science is Niklaus Wirth's *Algorithms Plus Data Structure Equals Programs* published in 1975. Wirth and other computer scientists conceptualized the intellectual work of programming as consisting of two interconnected parts: creating the data structures which logically fit with the task which needs to be done and are computationally efficient, and defining the algorithms which operate on these data structures.

We can use this conceptual model of computer programming to refine our understanding of what media applications do. All applications, including media software, are computer programs, so internally they involve algorithms working on data structures. This point by itself is not very revealing. What is more important is that these two elements, in my view, also define a mental model a user has of the application—i.e., how the user understands what an application presents to him/her, and what s/he is doing while using that application. In other words, the user's mental model reflects the abstract structure of a computer program (algorithms operating on data structure) that drives the particular media application software.

Inside the application environment, a user is working with one or more documents which contain content structured in a particular way. The user is aware of the importance of this structure—even though media applications do not use the term "data structure." The user understands that the data structure determines what content can be created and the operations which can be used to shape and modify it. If I choose vector graphics as my format, this implies that I will be creating straight and curved lines and gradients; I will be able to reshape any line in the future without losing quality; I will also be able to get perfect print output at any resolution. If I select bitmap image as my format, I can work with photographs and other continuous-tone images, blurring and sharpening details, painting over, applying filters and so on; but the price for this flexibility is that the image will exhibit undesirable artifacts if I enlarge it many times.[6] (In practice, this selection is done when a user chooses the primary application to assemble the project. Choosing Illustrator implies that you will be working with vector graphics; choosing Photoshop implies that you will be working with bitmap images. Although each program also supports working with the opposite image type, the majority of its functions and its interfaces are organized around its "native" type.)

---

[6] An example of a popular vector file format is Illustrator's AI format; JPEG and PSD are examples of bitmap file formats. Certain file formats such as EPS, PDF, and SWS can hold both vector graphics and bitmap images.

One more reminder because this is important: the term "data structures" has a particular meaning in computer science, referring to how data to be processed by a program is organized. As I have already said, while I want to retain the core idea of data organization, I am not interested here in how this actually happens on the low-level (i.e., whether the program organizes data using arrays, linked lists, etc.). Instead, I am using this term to refer to the *level of data organization, which is made visible and accessible to a user* and thus becomes a part of his/her mental model of the media creation and editing process. (For instance, when I work with bitmap images in Photoshop, I can zoom to examine individual pixels; I can check resolution of my image in pixels, I can choose the diameter of my brush, again in pixels; and so on. All this reminds me that my image is a pixel grid.)

To summarize this discussion: I suggested that both theoretically and also experientially—at least for the users who have more than casual experiences with media applications—"media" translates into two parts which work together. One part is a small number of basic data structures (or "formats") which are the foundation of all modern media software: bitmap image, vector image, 3D polygonal model, 3D NURBS model, ASCII text, HTML, XML, sound and video formats, KML, etc. The second part is the algorithms (we can also call them "operations," "tools" or "commands") that operate on these formats.

The ways in which these two parts are actualized in media applications require additional discussion. First, different applications often add more details on top on these basic types to give them additional functionality. For instance, an image as it is defined by capabilities of Photoshop (a professional and more expensive application) is substantially different from the way it is defined in the much less expensive Apple's iPhoto or Google's Picasa. As defined by Photoshop, an image is a complex hierarchical structure. (This description refers to Adobe Photoshop CS5.5.) At its base is the basic bitmap image: a grid of pixels. A pixel is a minimal element that a user can select and modify. This is a basic data structure Photoshop shares with all other image editors. (Note that while the user cannot select parts of pixels, the actual algorithms often work on a sub-pixel level.) A document in Photoshop can contain many such pixel grids; they are referred to as 'layers." The layers can be collected in groups. They can also form "layer comps"—alternative versions of the composition layout. Any single document layer can also have many states: a user can make it visible or invisible, change its transparency, the way it interacts with layers underneath, etc. Photoshop also provides special adjustment layers that do not contain any pixel content; according to the program documentation, "An adjustment layer applies color and tonal adjustments to your image without permanently changing pixel values."[7] Additionally, an image can also have a number of masks that define which areas of an image can be edited. Other elements that Photoshop adds to the basic image structure include paths and vector-based graphics and types.

Photoshop's UI uses a number of windows and menus to present this complex image structure, with all its possibilities. The Document window displays the actual composition. The Layers panel shows all layers, layer effects, and layer groups in the composition. The Channels panel shows the color components of an image (such as R, G, B). Each of these windows has a number of controls. Additionally, various menu items are dedicated to creating, viewing, and modifying all the possible image parts.

If data structures form one part of a user's mental model of media creation in software applications, the operations that can be used on these structures comprise the second part. That is, the user also understands that the process of defining and editing the content involves the sequential application of different operations provided by an application. Each operation corresponds to an algorithm which either performs some actions on the already existing data or generates new data. (Photoshop's Wave filter is an example of the former, while the Render Clouds filter is an example of the latter.)

In contemporary media software, the tools that are represented by items in menus are often referred to as "commands." The term "tool" is reserved for those frequently used operations that are given their own icons and can be selected directly without having to navigate through the menus. (I will use the word "tool" to refer to both types.) The applications group these related operations together. For example, Photoshop CS5 collects its key tools in the Tools Panel; additional tools are found under top-down menus

---

[7] http://help.adobe.com/en_US/Photoshop/11.0/

named Edit, Image, Layer, Select, Filter, and View. Many media applications also make available additional tools in the form of scripts that can be run from within the application. For example, in addition to dozens of commands already available in Photoshop CS5's menus and panels a user can also run Adobe or third-party scripts which appear under File > Scripts. These scripts can be written in JavaScript, VBScript, AppleScript, and other scripting languages. Finally, people who use command-line interfaces such as Unix (or Linux) can also use a third type of operations—separate software programs which are run directly from a command line. For example, two very widely used programs for image and video conversion and editing are ImageMagic and FFmpeg.[8] Since these types of programs do not have a GUI interface, they are not suitable for interactive image editing; however, they excel in automation and are often used to perform batch operations (such as conversion from one file format to another) on a large number of files at once.

Regardless of whether the media tools are presented via GUI, as scripts, or as separate programs available from a command line, they all have one thing in common—they can only work on particular data structures. For instance, image-editing applications define dozens of tools for editing bitmap images[9]—but these tools would not work on vector graphics. In another example, the techniques for modification of 3D models that define the volume of an object are different from the techniques that can operate on 3D models that represent the object's boundary (such as polygonal models).

To make an analogy with language, we can compare data structures to nouns and algorithms to verbs. To make an analogy with logic, we can compare them to subjects and predicates. Like all metaphors, these two highlight and distort, reveal and hide. However, I hope that they can help me to communicate my key point—the dual nature of a "medium" as defined by software applications.

We have now arrived at one possible answer to the question we posed in this book's introduction: what is media today as defined by software applications for its creation and editing? *As defined by application software and experienced by users, a "medium"*

[8] http://www.imagemagick.org/, http://ffmpeg.org/

[9] This Wikipedia article lists images editing operations common to these programs: http://en.wikipedia.org/wiki/Image_editing

is a pairing of a particular data structure and the algorithms for creation, editing and viewing the content stored in that structure.

Now that we have established that computational media involves combining algorithms and data structures, we can also better understand the distinction between media-specific and media-independent techniques that I made earlier. *A media-specific technique is an algorithm that can only operate on a particular data structure.* For example, blur and sharpen filters can only work on bitmapped images; the "extrude" operation commonly used in 3D programs to make 3D models can only be applied to a vector curve. In contrast, a *media-independent technique is a set of algorithms that all perform a conceptually similar task but are implemented to work on a number of data structures.* I mentioned examples of these techniques when I introduced this concept—sort, search, zoom, cut, copy, and paste, randomize, various file manipulations (copy, email, upload, compress, etc.), etc.

To explain how media-independent techniques can be implemented, let us look at the Copy, Cut and Paste commands. These operations already existed in some computer text editors in the 1960s. In 1974–1975 Larry Tesler implemented these commands in a text editor as part of Xerox PARC's work on a personal computer.[10] Recognizing that these commands can be used in all types of applications, the designers of Xerox Star (released in 1981) put dedicated keys for these commands in a special keypad.[11] The keypad contained keys marked Again, Find, Same, Open, Delete, Copy, Merge, and Move. A user could select any object in an application or on the desktop and then select one of these commands. Xerox PARC team called them "universal commands." Apple similarly made these commands available in all applications running under its unified GUI but got rid of the dedicated keys. [12] Instead, the commands were placed under the Edit pull-down menu.

The idea that a user can select objects in any document regardless of the media, or any file, and use the same set of commands on

---

[10] http://en.wikipedia.org/wiki/Cut,_copy,_and_paste

[11] http://en.wikipedia.org/wiki/Xerox_Star (February 20, 2012)

[12] For a close-up view of the dedicated keypad for universal commands and demonstration of their operations, see the part showing the Xerox Star keyboard in this video: http://www.youtube.com/watch?v=Cn4vC80Pv6Q&feature=relmfu (August 4, 2012).

these objects is among the most important inventions of the Xerox PARC team. It gives the user a single mental model of working with documents across applications, and simplifies learning new programs.

This is how the designers of Xerox Star described one of these universal commands:

> MOVE is the most powerful command in the system. It is used during text editing to rearrange letters in a word, words in a sentence, sentences in a paragraph, and paragraphs in a document. It is used during graphics editing to move picture elements, such as lines and rectangles, around in an illustration. It is used during formula editing to move mathematical structures, such as summations and integrals, around in an equation.[13]

However, depending on the type of media application and the kinds of objects a user selects, "copy," "cut" or "paste" will trigger different algorithms. For example, copying a phrase in a text document requires different sequences of operations than copying a selection in a bitmap image because the first is a one-dimensional sequence of characters, while the second is a set of pixels in a two-dimensional area. And even within a single application, many different algorithms will be needed to copy different kinds of objects a user can select.

My second example of how implementation of a media-independent technique involves different algorithms that work with particular media is the generation of random objects. The algorithm which generates a sequence of random numbers is very simple—it just calls the random number generator (a function available in all programming languages) to generate enough numbers, then scales these numbers within limits specified by the user (for instance, 0 to 1). This part is media-independent. Different applications can use this random number generation function as part of media-specific algorithms (i.e., algorithms which work on particular data structures) to create different types of content. For example, Photoshop has a command called "Add Noise" (located under Filters > Noise)

[13] D. Smith, C. Irby, R. Kimball, B. Verplank, B., E. Harslem, "Designing the Star User Interface," *Byte*, vol. 7, issue 4 (1982), p. 242–82.

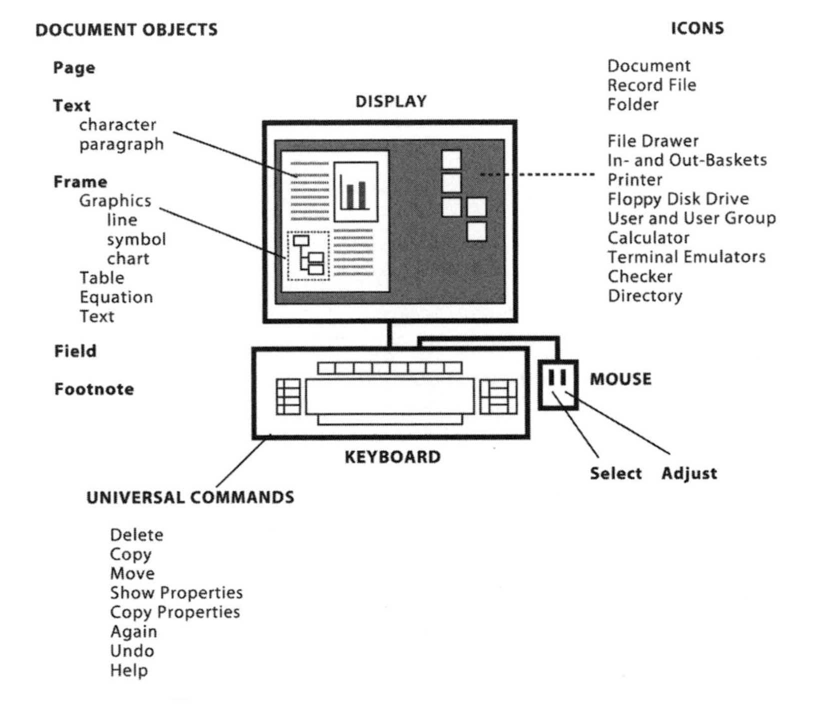

A diagram of the Xerox Star UI from D. Smith, C. Irby, R. Kimball, B. Verplank, B., E. Harslem, "Designing the Star User Interface," Byte, vol. 7, issue 4 (1982), 242–82. The universal commands are located in the dedicated keyword on the left part of the keyboard. (The original illustration from the article was redrawn in Illustrator.)

which generates a set of random X, Y number pairs. It then uses them to select specific pixels in the image and convert them to black or a random primary color (depending on the chosen option). A 3D modeling application can use the same technique to generate a set of identical 3D objects randomly located in space. Sound editing software can generate random sonic noise; and so on.

The implementation of media-independent techniques is structurally similar to various aesthetic systems in art that were not limited to a particular medium: for instance, baroque, neo-classicism, constructivism, post-modernism, and remix. Each system manifested itself across media. Thus, Baroque aesthetics can be found in architecture, sculpture, painting, and music; constructivism was applied to product design, graphic design, clothing, theatre, and possibly poetry and film.[14] But just as with media-independent techniques, realizing a particular aesthetic system in different media required some specific artistic devices that explored possibilities and worked with the limitations of each medium.

File formats

Software uses files to store and transfer data. For example, when you save a Photoshop image, all its channels, layers, groups, paths, and other information written to file using a particular format. Along with data structures, algorithms and UI, *a file format is* another fundamental element of computational media. File formats are the standardized mechanisms for storing and accessing data organized in a particular structure. Some file formats like .rdf are in public domain; some like .doc are proprietary. As I will discuss in more detail in the "Design Workflow" section in Chapter 5, standardization of file formats is an essential condition for interoperability between applications that in turn affects the aesthetics of media created with these applications. From the point of view of media and aesthetic theory, file formats constitute the "materiality" of computational media—because bits organized in these formats is what gets written to a storage media when a file is saved, and also

[14] Vlada Petric, *Constructivism in Film – A Cinematic Analysis: The Man with the Movie Camera* (Cambridge: Cambridge University Press, 1993).

because file formats are much more stable than other elements of computational media (I will explain this below).

Since both the materials and tools of physical media are now implemented as software, in theory new file formats and new algorithms can be easily created at any time, and existing ones can be extended. (Recall my discussion of "Permanent Extendibility" in Chapter 1.) However, in contrast to the 1960s and 1970s, when a few research groups were gradually inventing computational media, today software is a big global industry. This means that software innovation is driven by social and economic factors rather than by theoretical possibilities. As long as file formats are kept constant, it is easy to add new tools in subsequent software releases and the old tools will continue to work without modification. Moreover, in contrast to the period when Kay and others were defining "the first metamedium," today millions of individual professional users—as well as design and architecture firms, film studios, stock agencies, web design companies, and other creative companies and groups around the world—store their work and their assets (3D models, photographs, print layouts, websites, etc.) as digital files in particular formats: .doc, .pdf, tiff, .html, etc. If file formats were to change all the time, the value of media assets created or owned by an individual or a company would be threatened.

As a result, in practice file formats change relatively infrequently. For example, the JPEG image format has been in use since 1992, while the TIFF format goes back to 1986. In contrast, the modification of software tools which can work on files—creating, editing, displaying, and transmitting them—and the creation of new tools happens at a fast pace. When a company releases a new version of its application software, it usually adds various new tools and rewrites some of the existing ones but the file format stays the same. This stability of media file formats also allows other developers (both companies and individuals) to create new tools that work on these formats. In other words, it is one of the conditions that make "constant extendibility" of media software possible in practice. Here is an example of this extendibility: when I visited the plugins area of Adobe's Photoshop website on August 5, 2012 it listed 414 plugin products.[15] Given that a typical product can include a dozen or even

[15] http://www.adobe.com/cfusion/marketplace/index.cfm?event=marketplace.categories&marketplaceId=2&offeringtypeid=5 (August 5, 2012).

thousands of filters, presents, or action sets, the total number of available plugins is likely to run into the hundreds of thousands.

Each file format and its corresponding data structure has its strengths and weaknesses. A photograph represented as a bitmapped image can be given a painterly appearance; blurred and sharpened; composited with another photograph, and so on. All these operations are much more difficult or even impossible with a vector image. Conversely, it is much easier to edit complex curves if they are internally represented by mathematical formulas—a format used by vector drawing programs such as Illustrator and Inscape. Because many projects call for the combination of effects only possible with different data structures (such as raster and vector), over time professional software applications were extended to handle the corresponding file formats in addition to their native format type. For example, while the majority of Photoshop CS4 tools are geared towards raster images, it also includes some tools for working with vector drawings. Photoshop can also import vector graphics, while Illustrator can import bitmap images.

However, this hybridization of software applications does not change the fact that each separate application tool can only work on a particular data structure. This is true for universal commands such as "cut," "copy," "paste," and "view," as well as for a multitude of media-specific commands such as "word count," "blur," "extrude" and "echo." Thus, behind both media-independent and media-specific tools are separate algorithms each designed to work with particular data structures. However, a user has a different understanding of the two types, since the implementation is not visible directly. The former bring all media types together conceptually, and even creates an imaginary horizon where all differences between them disappear; at the same time, the latter emphasizes these differences since they only become available when a user works with a particular media.

As the name indicates, computer "files" refer to the paper files that were the key information management technology of a mid-twentieth-century office when computers were developed. The word "file" was used already in 1950 in RCA advertisement for its new "memory" vacuum tube; in 1952 the word was used to refer to information stored on punch cards.[16] With the devel-

[16] http://en.wikipedia.org/wiki/Computer_file

opment of the Web in the 1990s, web "documents"[17] such as web pages became equally important. A web page may consist of a single HTML file that contains static text, and other media content stored on a server. Alternatively, a web page may be "dynamic," which means that is constructed when the user accesses its address; it can also change as a result of user interaction.[18] Dynamic web pages can be constricted using client-side scripting (e.g. JavaScript); they can be also constructed using server-side scripting (PHP, Perl, Java, and other languages). The two methods can also be combined using Ajax techniques; for example, the popular web application that uses Ajax is Google Earth. In 2011, HTML 5, the next generation of the HTML standard, enabled multimedia and graphics elements including video, audio, and SVG graphics without the need for client side plugins. As these and other technologies were gradually developed and adopted, the identity of the web has been gradually changing—from static pages in the first half of the 1990s to "rich internet applications" that match much of the functionality of traditional desktop applications.[19]

Given the complexity and variety of web documents and applications types, the multitude of technologies and techniques for creating them, and the continuous evolution of both technologies and web conventions, it would not be appropriate to simply mechanically map our concept of media data structures to the web. If we instead focus more on the meaning of data structure as a mental model of media shared by designers and users, as opposed to its technical implementations, then this concept does apply to the web documents and applications. However, rather than referring to the type of media and its characteristics (text, bitmap image, vector image, sound, 3D model, etc.) that become elements of a web document or an application, we can also use it to describe the *interaction possibilities and conventions* offered by a web document or an application. For example, web pages typically continue to have hyperlinks that allow a user to go to related pages. Today a web page may contain "social media buttons" which allows users to easily share some content on the page. Particular genres of web documents and applications offer

[17] http://en.wikipedia.org/wiki/Web_document
[18] http://en.wikipedia.org/wiki/Dynamic_web_page
[19] http://en.wikipedia.org/wiki/Rich_Internet_application

their own interaction possibilities: for example, a blog typically contains a list of blog posts organized by dates; a webmail application contains buttons for responding, forwarding, and archiving email; and so on.

Although in principle I can go through all the most widely used types of web documents and applications, and write down a list of their conventions as they exist right now, this list will be both very long and no longer accurate by the time my book manuscript comes out in print. This is one of the reasons why this book focuses on discussion of media as *representation* rather than as *communication and interaction*—because the structures of software-based representations are more stable, less numerous, less complex, and change much less frequently than software and network based communication and interaction technologies. This does not mean that I give up on the project of *understanding web media* (including *mobile applications,* which currently number in the hundreds of thousands)—instead, I hope to do it more comprehensively in the future, while limiting myself to only brief discussions in this book.

Parameters

As I suggested earlier, a user's mental model of media creation and editing—both in the context of a particular application type, and when dealing with "digital media" in general—contains two fundamental elements, which, both conceptually and practically, correspond to two elements of computer programming: algorithms and data structures. When a user selects a particular tool from a menu and uses it on the part of the document s/he is working on, the algorithm behind this tool modifies the data structure that stores the content of the part. Most users, of course, do not know the details of how media software works on a programming level, and they may be only vaguely aware of what an algorithm is. (In 2011 I was driving towards San Francisco and saw a big ad board along the road, which prominently featured the word "algorithm"—but I could not stop to snap a picture that I really wanted to include in this book.) However, unknown to them, the principles of contemporary computer programming are "projected" to the UI level—shaping how users work with media via software applications practically, and how they understand this process cognitively. The data structure/algorithm model is one example of this. We will now look at another example of such "projection": options and their implementation.

One of the principles of modern computer programming—regardless of programming paradigm and programming language—is the use of parameters.[20] The popularity of parameters (which are also called "variables" or "arguments") is due to several reasons. One is the common modern programming practice of breaking a program into separate functions. If a program has to execute the same sequence of steps more than once, a programmer encapsulates this sequence in a single function which can be then evoked within the program by its name as often as needed. (Depending on the programming language, functions maybe be called procedures, methods, subroutines, or routines.) Dividing a large program into separate modular functions makes it easier to write, read, and maintain. (This programming paradigm is called Procedural Programming.) Functions that perform conceptually related tasks (for instance, generating graphics on the display screen) are collected in *software libraries*; such libraries are available for all popular programming languages and their use greatly speeds up software development. A function definition typically includes a number of parameters that control the details of its execution. For example, a function that translates an image into another format will have a parameter specifying whether the output format should be JPEG, PNG, TIFF, or another format. (And if you choose JPEG, you will get another parameter to specify the level of compression). The second reason for the popularity of parameters is that many functions (and whole programs) solve mathematical equations. A formula defines a relationship between variables. For example, a sine formula looks like this: y=A*sin(w*x + O), where A stands for amplitude, w is frequency, and O is phase. If we implement this formula as a software function, this function will have parameters for each variable (i.e., w, x and O).

If you do not program, you may still be familiar with the concept of parameters if you use formulas in Excel or Google Docs spreadsheet. For example, to generate a column of random

[20] http://en.wikipedia.org/wiki/Argument_(computer_science)

numbers, you can fill its cells with the function RAND(). The formula does not have any parameters; it simply generates random numbers that fall between 0 and 1. If you want to generate random numbers which fall within a different range, use a different formula RANDBETWEEN(bottom, top). This formula has two parameters which specify minimum and maximum values of the numbers of the range. For example, RANDBETWEEN(10, 20) generates a set of random values which lie between 10 and 20.

In a modern GUI the parameters, which control program execution, are called *options*. Users specify the values for the options using text boxes, sliders, buttons, drop-down lists, check boxes, and radio buttons. Sometimes the interface provides a few possible values and a user can only choose from them; in other cases she can enter her own value; in still other cases an application provides both possibilities. For example, a typical color picker allows a user to set R, G, and B values to create a unique color; alternatively s/he can choose from a set of predefined color swatches.

Use of options greatly expands the functionality of software applications by allowing the same application to perform a wider range of actions. For instance, imagine that you need to sort a set of numbers. Instead of using two separate programs to perform the sort in ascending and descending order, you can use a single program and set the parameter value, which would determine what kind of sort to perform. Or, imagine a round brush tool included in an image editing application's toolbox. You would not want to see a separate tool for every possible color, every possible brush radius, and every possible transparency setting. Instead, a single tool is given options which control color, radius, and transparency.

What does this mean for media theory? With softwarization, the implicit possibilities and different ways of using physical tools are made fully explicit. "Artistic techniques" and "means of expression" (see the definition of a medium which opens Part 1) are given explicit and detailed controls. Like all computer programs or functions, they now come with many parameters. For instance, Photoshop CS5's basic brush tool has these parameters: size, hardness, mode, airbrush capacity, opacity, and flow. Hardness, opacity, and flow can have any value between 0 and 100; there are 25 modes to choose from; and the diameter can vary from 1 pixel to 2500 pixels.

Do most users need all these options and such a degree of precision? Probably not. For example, controlling opacity in 5 percent intervals will probably be quite sufficient for most users. However, the algorithm that implements opacity behavior is exactly the same regardless of whether a particular parameter can have 10 settings, or 100. Since the general logic of software industry is to always try to offer users "more" than the previous application's version or competitors, it is understandable that the brush interface gives us the larger number of options and a larger choice of their values, even though such precision may be not needed. The same goes for most other tools available in media software. In this way, the logic of programming is projected to the GUI level and becomes part of user's cognitive model of working with media inside applications.

While adding more options does require additional programming labor, offering more values for these options typically does not. If I want the brush tool to have a transparency option, I need to write new code to simulate this behavior. However, giving a user a choice of 20 or 100 possible values for transparency does not cost me anything. Like any other media tool a brush tool is an algorithm that takes some inputs and generates outputs by applying a number of computations to these inputs. In the case of a brush tool, the inputs are the values of the options set by the users, and the colors of the pixels over which the brush moves; the outputs are the new pixel values. The algorithm does not care what the particular input values are, and the number of steps required to execute them also does not change because of particular values.

Pre-industrial physical media tools did not have explicit controls. There were no numerical parameters to set on a pen, brush, or chisel. If you wanted to change the diameter of a brush, you picked a different brush. The industrial era introduced new types of media tools that were mechanical or electronic machines: the electrical telegraph, photo camera, film camera, film projector, gramophone, telephone, television, and video camera. Like all industrial machines they now came with a few controls. Physically these controls appeared as knobs, levers, and dials. The next generation of media—*software applications*—give explicit controls to all the tools they include. The tools that did not have explicit controls before now acquired them; the tools that had a few were given many more. (While machines made from mechanical parts can have only a limited number of possible settings, software parameters typically can have practically unlimited range of values.)

The Language of New Media introduced the idea of "transcoding"—the mapping of the conventions and principles of software engineering to cultural concepts and perceptions. The explicit *parameterization* of all media creating and editing techniques implemented in software is a perfect instance of transcoding logic. Thus, the fact that Photoshop's brush tool comes with a number of options controls and that opacity and flow can take any value between 0 and 100 is only partially related to the meaning of this command—a simulation of different physical brushes. The real reason for this implementation of the command lies in its identity as a computer program. (As I already discussed, in modern programming, programs and their parts are given parameters, which in many cases can take any arbitrary numbers as inputs.)

Along with being a good example of how principles and conventions of software development in general are carried over into media applications and our media lives, parameterization also exemplifies another trend. Seemingly different media activities—editing photographs, creating 3D game characters, editing a video, or working on a website design or a mobile application—become similar in their logic and workflow: select a tool, choose its parameters, apply it; and repeat this sequence until the project is done.

Of course, we should not forget that the practices of computer programming are embedded within the economic and social structures of the software and consumer electronics industries. These structures impose their own set of constraints and prerogatives on the implementation of hardware and software controls, options, and preferences (all these terms are just different manifestations of software parameters). Looking at the history of media applications and media electronics devices, we can observe a number of trends. Firstly, the number of options in media software tools and devices marketed to professionals gradually increases. For example, a significant number of Photoshop's tools and filters have more options and controls than in earlier versions of the application. Secondly, features originally made available in professional products later become available in consumer-level products. However, to preserve the products' identities and justify price differences between different products, a significant difference in feature sets is continuously maintained between the two types of products, with professional software and equipment having more tools and more parameters than their consumer equivalents. Thus Photoshop has lots of tools, Photoshop Elements offers fewer tools and iPhoto and Picasa have fewer still.

All this is obvious—however, there is also a third trend, more interesting for media theory. Following the paradigm already established by the end of the nineteenth century when the Kodak company started to market its cameras accompanied by a slogan "you push the button, we do the rest" (1892), contemporary software applications and media devices aimed at consumers significantly *automate media capture and editing* in comparison to their professional counterparts. For example, during the 2000s many consumer digital cameras only offered automatic exposure; to get full manual controls one had to go to the next price category of semi-professional cameras. In another example, towards the end of that decade, consumer cameras started to incorporate automatic face and smile detection—features that were not available on expensive professional cameras.

Like any other type of automation, automatic exposure requires more computing steps than the use of manual settings; the same goes for applying "auto contrast" or "auto tone" commands available in media software. Thus, if we equate the use of computers with automation, paradoxically it is the consumers who fully enjoy its benefits—in contrast to professionals who have to labor over all these manual settings of all these controls... but of course, this is what they paid for: to achieve effects and results that built-in automation cannot deliver. At the same time, by offering more high-level automation in consumer-priced products, the industry unintentionally undermines the professionals' skills. For instance, today a number of web-based applications and application plug-ins can take a portrait photo and improve contrast, correct skin tone, remove skin imperfections and wrinkles—all in one step. The results can be surprisingly good—maybe not good enough for the cover of *Vogue* but quite sufficient for posting the improved photo to one's profile on a social network site.

Interestingly, in the beginning of the 2010s this trend was partly reversed. Because of the much larger size of the consumer market and faster release cycles, hardware and software makers started to offer some new features first in lower-level products and then only later in more expensive products. For example, Apple's Aperture 3 (2010) added options which had already been made available earlier in a consumer-level iPhoto application in its 2009 release—Faces (face recognition) and Places (a system for identifying geo locations of the photos and locating them on a map interface).[21]

The metamedium or the monomedium?

Given fundamental similarity in how "mediums" function as implemented in software, a logical question arises: do we need to talk about different mediums at all? In other words, is the computer metamedium a collection of simulated, new, and yet-to-be-invented media, or is it one "mono-medium"? Are we dealing with *the metamedium or the monomedium?*

We now understand that in software culture, what we identify by conceptual inertia as "properties" of different mediums are actually *the properties of media software*—their interfaces, the tools, and the techniques they make possible for accessing, navigating, creating, modifying, publishing, and sharing media documents. For example, the ability to automatically switch between different views of a document in Acrobat Reader or Microsoft Word is not a property of "electronic documents" in general, but as a result of software techniques whose heritage can be traced to Engelbart's "view control." Similarly, the ability to change the number of segments that make up a vector curve is not a property of "vector images"—it is an option available in some (but not all) vector drawing software.

As we have learned, every media software also includes at least some tools that are not media-specific—i.e. they are not limited to working on particular data structures like raster images or vector drawings. Originally conceptualized at Xerox PARC as ways to enable users to transfer the cognitive habits learned in using one application to other applications, today a small number of Xerox's "universal commands" have become a larger number of

[21] http://photo.net/equipment/software/aperture-3/review/ (March 4, 2012).

"media-independent" media tools and interface techniques which are shaping how users understand all types of media content.

Despite these fundamental ways in which distinct mediums become aligned conceptually and practically, I do not want to abandon the concept of different mediums altogether. The substantial differences between the operations for media authoring and editing supported by different data structures (in the sense of this term used here) is one reason to keep this concept. And here are three additional reasons:

1. "Mediums" as they are implemented in software are part of distinct cultural histories that go back for hundreds and often thousands of years. Electronic text is part of the history of writing; digital moving images are part of the history of a moving image which includes shadow plays, phantasmagoria, nineteenth century optical toys, cinema, and animation; a digital photograph is part of almost 200 years of photography history. These histories influence how we understand and use these media today.

Put differently, we can say that any film today exists against the horizon of all films ever made and a subset of those films which a particular person making or watching a particular film has seen in their life; similarly, any digital image exists against the horizon of all images in a "museum without walls" (André Malraux) of human visual history. A medium, then, is not just a set of materials and tools (whether physical, mechanical, electronic, or implemented in software) and artistic techniques supported by these tools—it is also an imaginary *database of all expressive possibilities, compositions, emotional states and dynamics, representational and communication techniques, and "content" actualized in all the works created with a particular combination of certain materials and tools.*

Systematic digitization of cultural heritage is gradually turning this imaginary database into a real one. By 2012, Google has digitized 20 million books, while artstor.org offered over one million images of art and architecture digitized by 228 museums and private collections. However, this process did not start with digital computers. In the early twentieth century the development of public art museums, illustrated art magazines and books, lantern slide lectures, and the academic study of arts in the universities made large numbers of previously created artworks directly visible and accessible to the public (as opposed to small numbers only accessible to art patrons). For instance, American museums started to develop lantern slide collections of their holdings after 1865; in 1905 the University of California at Berkeley offered the first architecture history course that used lantern slides.[22]

The digitization of cultural collections since 1990 started to gradually bring together the materials dispersed among all these already available resources, making them searchable and accessible through single websites. For example, by early 2012 Europeana[23] provided information and links to 20 million digitized cultural objects including paintings, drawings, maps, books, newspapers, diaries, music and spoken word from cylinders, tapes, discs and radio broadcasts, and films, newsreels and TV broadcasts, contributed by 1500 European institutions.[24] In the UK, the BBC's Your Painting project was set up to offer free online access to digital images of all 212,000 paintings from all UK national collections; this goal was achieved by 2013.[25] In the US, the Library of Congress provides access to dozens of digital collections from a single portal. The collections include 4.7 million high-resolution newspaper pages (1860–) and over one million digital images, such as 171,000 scanned photo negatives from the Farm Security Administration/Office of War Information program (1935–45).[26]

These institutional digitized collections are supplemented by user-uploaded born-digital and digitized cultural artifacts. YouTube and other video sharing sites contain a substantial sample of all cinema history in the form of short clips. Flickr has a large number of photos of artworks taken by museum visitors around the world. Portfolio sites for professional media creators such as Coroflot.com and Behance.com contain millions of portfolios in art direction, exhibition design, illustration, interaction design, motion graphics, and other fields. Manga scanlation websites contain millions of fan scanned and translated manga pages (as of March 4, 2012, mangapark.com hosted 5,730,252 pages for 9020 manga series[27]); Scribd.com hosts tens of millions of text documents (you will

[22] http://en.wikipedia.org/wiki/Slide_library
[23] http://www.europeana.eu/
[24] http://pro.europeana.eu/web/guest/news/press-releases (March 4, 2012).
[25] http://www.bbc.co.uk/arts/yourpaintings/ (January 31, 2013).
[26] http://www.loc.gov/library/libarch-digital.html (March 4, 2012).
[27] http://www.mangapark.com/ (March 4, 2012).

most likely find this book there as well); the deviantArt online community for user-generated art hosts over 100 million submissions from 190 countries (Spring 2012 data).[28]

Although the distribution of what is available in all these online archives is highly uneven in relation to the types of media, historical periods, countries, and so on, it is safe to say that today the cultural classics—i.e. works of famous film directors, cinematographers, graphic designers, media designers, composers, painters, writers, and so on—are all available online either in complete form or in parts. As a result, the idea of a "medium" in a sense of *all the creative works and possibilities realized so far using a set of technologies* has become quite real. As opposed to holding the imaginary database of key works in a particular medium in your head (not the most reliable place for such large-scale storage), you can now quickly consult the web to locate and study any of them. Further, you are not limited to the holdings of a single museum or a library at a time; instead, you can use Europeana, Artstor, or other large-scale collection aggregators to browse their combined metacollections. For example, while the National Gallery in London has 2,300 paintings, Your Painting site from BBC offers images of 212,000 paintings from all UK national collections.

2. We can also use the term "medium" to refer to a *presentation/interaction platform*. If you think of iOS and Android platforms (each containing mobile devices, an operating system, and apps) as examples, you are right—but I also would like to use the word "platform" in a more general sense. Medium as a *platform* refers to a set of resources that allows *users* to access and manipulate content in particular ways. Understood in that way, a white cube in a modern art gallery is a medium; so are a modern movie theatre, a print magazine, network television, and a DVD. (Note that just as iOS and Android are ecosystems that combine many elements, many older media platform work similarly. Movie theatres, film production, distribution, and publicity form the cinema platform; so do television production, distribution, and TV sets.)

This meaning of "medium" is related to the twentieth-century concept of "media" in Communication Studies, which came from Shannon's Information Theory: the storage and transmission

[28] http://en.wikipedia.org/wiki/deviantArt

channels and tools used to store and deliver information. However, if the latter concept includes both storage and transmission channels and tools, my "presentation platform" puts more focus on the reception technologies. These technologies may include special spaces, architecture (for example, large media surfaces which are increasingly becoming part of buildings' interiors and exteriors), sensors, lights, and, of course, media viewing/editing/sharing devices and apps. Presentation platforms also "program" particular patterns of behavior: one walks through and touches architecture; stays silent in a movie theatre; interacts with family members while watching TV, moves one's body in front of an interactive wall; etc.

During the nineteenth and twentieth centuries, presentation platforms were closely tied up with particular types of media content. Art museums displayed paintings and sculptures (and later performances and installations); newspapers published texts and images; TV presented television shows, news programs, and films. The gradual addition of media viewing and playing capabilities to computers (and later laptops, mobile phones, tablets, media players and other computer-based devices) loosened this connection. Distribution, storage, and presentation of different media types were no longer tied up to particular technologies and particular presentation platform. (This separation parallels the similar process of "softwarization" I have discussed in detail: the tools and techniques for media authoring and editing have become liberated from their reliance on particular physical and electronic technologies.)

Today, we can access most types of media on every computer platform. You can view images, videos, text documents and maps inside email, in a browser, on your notebook, a PC, laptop, tablet, mobile phone, internet-enabled TV, or an in-car or in-flight entertainment system. What differentiates these devices is neither the types of contents they can play, nor the basic interfaces they provide for viewing and interacting with content, but rather the relative ease with which one can navigate various media.

For instance, my 2011 Samsung LCD-TV came with a full web browser—however it was much better at playing cable TV and Netflix content than web browsing experience, which was quite hard with a TV remote control. In another example, the relatively small screen of a mobile phone in the early 2010s, its less powerful processor, and its smaller amount of RAM makes it a less-than-ideal platform for editing feature films or doing CAD (Computer Aided Design). At the same time, the same small size gives a mobile phone platform many advantages. In many countries it is socially acceptable for a person to engage in chat or send and receive SMS on their phone during a social meal—but doing it on a laptop would be inappropriate. The size of a mobile phone also makes it a perfect device for location-based social networking (e.g., Foursquare) and other services: recommending social events in a city, following friends on a map, playing location-aware games, and so on.[29]

Similar to mobile phones, many platforms take advantage of their social settings by adding unique features. For example, some airline in-flight entertainment electronics systems (e.g., Virgin America's and V Australia's *RED* Entertainment System[30]) allow passengers to chat with other passengers or participate in multi-player games with them.

As a media presentation and interaction platform, each type of computer-based consumer device has its differences. Currently (early 2013), some mobile platforms such as iOS do not allow users to save documents directly to the device; instead, files can only be saved inside their respective apps. [31] Some media devices such as e-book readers, video players, audio players, digital billboards, and game consoles often can only play a few (or even just one) type of media content. (While consumer electronics companies are engaged in the ongoing "convergence" war, gradually adding ability to play all media types to most devices, this trend does not affect every single device.)

Given the differences in their physical appearance (size, weight, form factor), physical interface (touchscreen, keyboard, remote control, voice input, motion sensing) and media playback/editing/sharing/networking capabilities, it is tempting to think of each type of device as a different "medium." As presentation/interaction platforms these devices provide distinct user experiences and encourage distinct sets of media behaviors (sharing location, working, chatting, etc.). However, we should also remember that since they all use the same technologies—computers, software, and networks—they also share many fundamental features.

---

[29] http://en.wikipedia.org/wiki/Location-based_service

[30] http://en.wikipedia.org/wiki/In-flight_entertainment

[31] http://en.wikipedia.org/wiki/Mobile_platform

Given this, we may want to think of all these presentation/interaction platforms using the idea of "family resemblance" articulated by many nineteenth-century thinkers and later by Wittgenstein—that things may be "connected by a series of overlapping similarities, where no one feature is common to all."[32] But this would not be accurate. Prototype theory, developed by the 1970s by psychologist Eleanor Rosch and other researchers, may fare better. Based on psychological experiments, Rosch showed that for a human mind, some members of many semantic categories are better representatives of these categories than other members (or example, a chair is more prototypical of the category "furniture" than a mirror.[33])

If we consider the best current implementation of Kay's vision augmented with networking and sharing capabilities as the prototype (i.e. the most central member of the category) "of the computer metamedium" (for me, my current Apple laptop would qualify as such a prototype, but any full-featured laptop will also do), then all other computer devices can be situated at some distance from the prototype based on how well they instantiate this vision. Of course, it is also possible to argue that none of the current computers or computer devices realizes Kay's vision sufficiently, because casual users cannot easily program them and invent new media. (Appropriately, Kay named his 1997 Turing Award lecture "The Computer Revolution Hasn't Happened Yet."[34]) In this interpretation, Kay's Dynabook is the imaginary ideal prototype, and each realized computer device is situated at some distance from it.

3. Another important meaning of the concept "medium" is related to human sensory systems, which acquire and process information in different ways. Each sensory system contains sensory receptors, neural pathways, and particular parts of the brain responsible for further processing. Traditional human cultures recognized five senses: sight, hearing, taste, smell, and touch. Additionally, humans can also sense temperature, pain, positions of body parts, balance, and acceleration.

---

[32] http://en.wikipedia.org/wiki/Family_resemblance

[33] Eleanor Rosch, "Cognitive Representation of Semantic Categories." *Journal of Experimental Psychology: General* 104, no.3, (September 1975): pp. 192–233.

[34] http://blog.moryton.net/2007/12/computer-revolution-hasnt-happened-yet.html (March 5, 2012).

Because the concept of senses has been important for ancient and modern Western philosophy, Buddhist thought, and other intellectual traditions, the discussions of senses in relation to art and aesthetics have also been very extensive, so to enter them seriously would require its own book. Instead, I will limit myself to a short discussion of a more recent research in cognitive psychology that can be used to support the idea of multiple mediums: how different types of information are represented and manipulated in the brain.

The important question debated in cognitive psychology for forty years is whether human cognition operates on more than one type of mental representation. One view is that the brain only uses a single propositional representation for different types of information. According to this view, what we experience as mental images are internally translated by the mind into language-like propositions.[35] The alternative view is that the brain represents and processes images using a separate representational system. After decades of psychological and neurological studies, the current consensus supports this second view.[36] Thus, it is believed the brain operates on and maintains mental images as mental image-like wholes, as opposed to translating them into propositions.

If we accept this view that language, and images/spatial forms require the use of different *mental processes and representations* for their processing—propositional (i.e., concept-based, or linguistic) for the former, and visual/spatial (or pictorial) for the latter, it helps us to understand why the human species needed both writing and visual/spatial media. Different mediums allow us to use these different mental processes. In other words, if media are "tools for thought" (to quote again the title of a 1984 book by Howard Rheingold about computers[37]) through which we think and communicate the results of our thinking to others, it is logical that we would want to use the tools to let us think verbally, visually, and spatially.

---

[35] http://plato.stanford.edu/entries/mental-representation/

[36] Tim Rohrer, "The Body in Space: Dimensions of Embodiment," in *Body, Language and Mind*, vol. 2, eds. J. Zlatev, T. Ziemke, R. Frank, R. Dirven (Berlin: Mouton de Gruyter, 2007).

[37] Howard Rheingold, *Tools for Thought: The History and Future of Mind-Expanding Technology*, a revised edition of the original book published in 1985 (The MIT Press, 2000).

An hypothesis about the existence of propositional and pictorial representations in the mind is one example of a number of theories which all share the basic belief that human thinking and understanding are not limited to the use of language. For example, in 1983 Howard Gardner proposed a theory of multiple intelligences, which included eight different categories: bodily-kinesthetic, verbal-linguistic, visual-spatial, musical, interpersonal, logical-mathematical, naturalistic, and intrapersonal. Recall also that Alan Kay based the design of the GUI on the work of the psychologist Jerome Bruner who postulated the existence of three modes of representation and cognition: enactive (action-based), iconic (image-based) and symbolic (language). And while in Xerox PARC's implementation the appeal to the first modality was limited to the selection of objects on the screen using a mouse, more recently they were joined by touch interfaces and gesture interfaces. Thus, the interfaces of modern computers and computer-based devices are themselves prime examples of a multiple media at work—adding over time more media as interaction mechanisms, rather than converging to a single one, like written language (original UNIX and other OS from 1960s–1970s) or speech (Hal from Kubrick's *2001*).

The evolution of media species

As we see, while softwarization redefines what mediums are and how they interact, it does not erase the idea of multiple distinct mediums. In fact, in contrast to the idea of "convergence" popular in the 2000s in discussing the coming together of computers, television, and telephony, I would like you to think of the computational media using the concept of biological evolution (which implies increasing diversity over time). And here comes the ultimate difficulty with continuing to use the term "medium" as a useful descriptor for a set of cultural and artistic activities. The problem is not that multiple mediums converge into one "monomedium"—they do not. The problem is exactly the opposite: they multiply to such extent that the term loses its usefulness. Most large art museums and art schools usually have between four and six departments which supposedly correspond to different mediums (for example, San Francisco's Museum of Modern Art divides its collection into painting and sculpture, photography, architecture and design, and media arts[38])—and this is OK. We can still use unique names for different mediums if we increase their number to a couple of dozens. But what to do if the number goes into thousands and tens of thousands?

And yet this is exactly the situation we are in today because of softwarization. Extrapolating from the dictionary definition of a medium that opens Part 1, we can say that different mediums have sufficiently dissimilar representational, expressive, interactive and/or communicative capacities. (How much of a difference is needed for us to declare that we have two mediums rather than one? This is another fundamental question that makes it challenging to use the term "medium" in software culture.) Consider all the factors involved in gradual and systematic expansion of these capacities: "permanent extendibility" of software; the development of new types of computer-based and network enabled media devices (game platforms, mobile phones, cameras, e-book readers, media players, GPS units, digital frames, etc.) and the processes of media hybridization manifested in software applications, technology prototypes, commercial and artistic projects... Do we get a new medium every time a new representational, expressive, interaction or communication functionality is added, or is a new combination of already existing functions created? For example, does the addition of voice interface to a mobile device create a new medium? What about an innovative combination of different visual techniques in a particular music video? Does this music video define a new medium? Or what about the versions of Google Earth that run on a computer and support different types of layers, and the iPhone version which in 2008 only supported Wikipedia and Panoramio layers? Are these two different mediums?

If we continue to hold to our extended medium definition (extrapolated from the dictionary definition), we would have to answer yes to all these questions. Clearly, we do not want to do this. Which means that the conceptual foundation of media discourse (and media studies)—the idea that we can name a relatively small

[38] http://www.sfmoma.org/explore/collection (March 6, 2012).

number of distinct mediums—does not hold anymore. We need something else instead.

To explain this in an alternative way, when we considered only one aspect of media ecosystem—media design—we were able to adopt the traditional understanding of "medium" (materials + tools) to describe the operations and interfaces of application software. We did this by proposing this new definition: *medium = algorithms + a data structure*. Following this perspective, we can refer to simple text, formatted text, vector graphics, bitmap images, polygonal 3D models, spline-based models, voxel models, wave audio files, MIDI files, etc. as separate mediums. But even with this approach it was already difficult to cover the design of interactive applications and websites. When we start considering the larger ecosystem of the proliferating devices, network services, interface technologies, media projects, and over one million apps for mobile platforms available to consumers, the concept can no longer be stretched to describe them in a meaningful way. In this section, I will explore one way to think beyond "medium."

As I suggested above, the model of multiple species related via evolutionary development that we can borrow from biology offers a plausible alternative. This model is useful for thinking about both media authoring/editing applications and particular media projects/products (which, after all, are also software applications but more specialized—if media applications are content agnostic, projects typically offer particular content). The key advantages of a "species model" over a "mediums model" are their *large numbers* (Earth contains many million species—at least for now); their *genetic links* (which implies significant overlap in features between related species); and the concept of *evolution* (which implies constant development over time and gradually increasing diversity).

Each of these advantages is equally important. Instead of trying to divide the extremely diverse media products of software culture into a small number of categories (e.g. "mediums"), we can think of each distinct combination of a particular subset of all available techniques as a unique "media species." Of course, a software application or a project/product is not limited to remixing already existing techniques; it can also introduce new one(s), which may then reappear in other applications/products. Such new introductions can perhaps be thought as new genes (keeping in mind all the limitations of our use of these biological metaphors). If contemporary sciences such as evolutionary biology, genetics and neuroscience can describe and map (or at least, work towards describing and mapping) millions of distinct species, 3 billions of DNA pairs in the human genome, and 100 billion neurons in a cerebral cortex, why cannot media theory and software studies deal with the diversity and variability of software culture by providing richer classifications than the kinds we currently use? (For inspirational examples from live sciences, see the Human Connectome Project to create a comprehensive map of the adult brain,[39] and even more ambitious, The Blue Brain Project to create a completely realistic software simulation of the brain, accurate on a molecular level.[40])

Equally valuable is the notion that related species share many features. Eighteenth- to twentieth-century aesthetic tradition, from Gotthold Lessing to Clement Greenberg, repeatedly insisted on opposing a small number of mediums to each other, thus seeing them as distinct and non-overlapping categories. This trend intensified in European modernism of the 1910s–1920s, when artists tried to reduce every medium to its unique qualities. To do so, they gave up representation and concentrated on the material elements thought to be unique to each medium. Poets, such as Russian futurists, were experimenting with sounds; filmmakers proposed that the essence of cinema was movement and temporal rhythm (French film theory of the 1920s[41]) or montage (Kuleshov's group in Russia); and painters were exploring pure colors and geometric forms. For example, one of the pioneers of abstract art, painter Wassily Kandinsky, published two articles in 1919 with the titles which clearly signal his program to articulate an artistic language consisting only of basic geometric elements: "Small Articles About Big Questions. I. About Point," and "II. About Line."[42]

Although artists in the second half of the twentieth century systematically revolted against this trend, instead embracing

---

[39] http://www.humanconnectomeproject.org/about/ (March 7, 2012).

[40] http://bluebrain.epfl.ch/ (March 7, 2012).

[41] See Jean Epstein, "On Certain Characteristics of *Photogénie*," in *French Film Theory and Criticism*, ed. Richard Abel (Princeton: University of Princeton Press, 1988), 1: pp. 314–18; Germaine Dulac, "Aesthetics, Obstacles, Integral *Cinégraphie*," in *French Film Theory and Criticism*, 1: 389–97.

[42] Kandinsky further developed the ideas in these articles in his 1926 book *Point and Line to Plane*.

"mixed media," "assemblage," "multimedia art," and "installations," this did not lead to a new, richer system for understanding art. Instead, only a few new categories were created (e.g., the ones I just listed), and all the extremely diverse new art objects created by these artists were placed in these categories.

Evolutionary biology instead gives us a model of the much larger space of objects, which overlap in their identities. This model fits much better with my theory of software culture as a large and continuously growing pool of techniques that can enter into numerous combinations, in the form of applications, and projects/products created with them, or through custom programming. Obviously, the majority of these "media species" share at least some techniques. For example, all applications with a GUI interface designed to run on a full-size computer screen will use similar interaction techniques. To take another set of media species I have already analyzed as examples of media hybridity, Fujihata's *Alsace,* Sauter's *Invisible Shape,* and Google Earth all share a technique of embedding videos inside a virtual navigable space. Both Fujihata and Sauter construct unique interfaces to video objects in their 3D worlds. In contrast, Google Earth "inhabits" the YouTube video interface when you embed video from YouTube in a placemark (although you have some control over the embedded player characteristics[43]). This ability to embed the YouTube video player offers yet another example of a technique now used in numerous websites and blogs; so is the use of APIs offered by most major social media services and media sharing sites.

Finally, another attraction of the "species model" is the idea of evolution—not the actual mechanisms of biological evolution on Earth as theorized and argued over by scientists (because these mechanisms do not fit technological and cultural evolution), but rather the image of gradual but continuous temporal development and increased variability and *speciation* (emergence of new species) this idea implies—without implying progress. Without subscribing to the theory of memes, we can name a number of ways in which new techniques are transmitted in software culture: new projects and products that are seen by other designers and programmers; scientific papers in computer science, information science, HCI,

---

[43] YouTube Help, "Embedded player style guide," http://support.google.com/youtube/bin/answer.py?hl=en&answer=178264 (March 7, 2012).

media computing, visualization and other related fields; and also the code itself. This last mechanism is particularly important for us because, in contrast to previous media, it is unique to software culture (at the same time, it has a direct parallel with genetic code). As new techniques for media editing, analysis, interaction, transmission, retrieval, visualization, etc. become popular, they are coded in multiple programming languages and scripting environments (Java, C++, JavaScript, Python, Matlab, Processing, etc.) and become available as either commercial or, increasingly, open source software libraries ready for download online. If a programmer working on a new application, website, or any other media project wants to use any of these techniques, s/he can include the necessary functions in his/her own code.

Traditionally, new cultural techniques were transmitted via imitation and training. Professional artists or artisans saw something new and they imitated it; pupils, assistants, or students would learn techniques from their teachers; art students would spend years making copies of famous works. In either case, the techniques moved from mind to mind and from hand to hand. Modern media technologies add a new mechanism for cultural transmission—the interfaces and manuals of media devices, which carry with them the suggested, proper, ways of using these devices. All these mechanisms are at work today as well (of course, speeded up by the web). However, they are also joined by a new one—*the transmission of cultural techniques via algorithms and software libraries*.[44] This does not necessarily mean that such transmission does not introduce changes—programmers can always modify the existing code to the needs of their project. More important I think is the fundamental *modularity* of cultural objects created via this mechanism. On the level of techniques, a cultural object becomes an agglomeration of functions drawn from software libraries—the DNA it shares with many other objects that use the same techniques. (While commercial media products also widely use content elements such as photos purchased from stock libraries, this kind of modularity

---

[44] Jeremy Douglass suggested that we can study the propagation of techniques across software culture by tracking the use of particular software libraries and their functions across programs. Jeremy Douglass, presentation at SoftWhere 2008 workshop, University of California, San Diego, May 2008, http://workshop.softwarestudies.com/

is not openly acknowledged.) Because techniques are coded as software functions, this cultural modularity is closely linked to the principle of modularity in modern computer programming—the concept that a program should contain a number of self-contained parts. If a media project or an app introduces a new technique (or techniques) and they appear to be valuable, often the programmers make them available as stand-alone functions which then enter the overall pool of all techniques available within the computer metamedium—and thus easing the ways for others to adopt this technique in creating new projects.

I would like to end this chapter by quoting the historian Louis Menand who explains that while prior to Darwin scientists viewed species as ideal types, Darwin shifted the focus to variation—which for me is an overarching reason for thinking about media in the software age through the terms of evolutionary biology:

> Once our attention is redirected to the individual, we need another way of making generalizations. We are no longer interested in the conformity of an individual to an ideal type; we are now interested in the relation of an individual to the other individuals with which it interacts... Relations will be more important than categories; functions, which are variable, will be more important than purposes; transitions will be more important than boundaries; sequences will be more important than hierarchies.[45]

[45] Louis Menand. *The Metaphysical Club* (New York: Farrar, Straus, and Giroux, 2001), p. 123.

PART THREE

Software in action

CHAPTER FIVE

Media design

"We shape our tools and thereafter our tools shape us."
Marshall McLuhan, *Understanding Media* (1964)

After Effects and the invisible revolution

Media hybrids are not limited to particular software applications, user interfaces, artistic projects, or websites. If I am right in suggesting that hybridity represents the next logical stage in the development of computational media, following the first stage of simulating individual physical media in a computer, then we can expect to find it in many cultural areas. And this is indeed the case. In this chapter I will look at a single cultural area in depth—moving image design—analyzing how the creation and aesthetics of moving images changed dramatically in the 1990s.

Around the middle of the 1990s, the simulated physical media for moving and still image production (cinematography, animation, graphic design, typography), new computer media (3D animation), and new computer techniques (compositing, multiple levels of transparency) met within a single software environment—compatible software programs running on a personal workstation or a personal computer. Filmmakers, animators, and designers started to systematically work in this environment, using software both to generate individual elements and to assemble all elements together. The result was the emergence of a new visual language that quickly became the norm.

Today this language dominates the visual media produced in dozens of countries. We see it daily in commercials, music videos, TV graphics, film titles, interactive interfaces of mobile phone and other devices, dynamic menus, animated web pages, graphics for mobile media content, and other types of animated, short non-narrative films and moving-image sequences being produced around the world by media professionals including companies, individual designers and artists, and students. All in all, I estimate that at least 50 percent of short moving image works follow this language. In this chapter I will analyze what I perceive to be some of its defining features: new hybrid visual aesthetics; systematic integration of previously non-compatible media techniques; use of 3D space as a platform for media design; constant change on every visual dimension; and amplification of cinematographic techniques.

The new hybrid aesthetics exist in endless variations but its basic principle is the same: juxtaposing previously distinct visual aesthetics of different media within the same image. This is an example of how the logic of media hybridity restructures a large part of culture as a whole. The languages of design, typography, animation, painting, and cinematography meet within the computer. Therefore, along with being a metamedium as formulated by Kay, we can also call a computer a *metalanguage* platform: the place where many cultural languages of the modern period come together and begin creating new hybrids.

How did this language come about? I believe that looking at the software involved in the production of moving images goes a long way towards explaining why they now look the way they do. Without such analysis we will never be able to move beyond the commonplace generalities about contemporary culture—post-modern, global, remix, etc.—to actually describe the particular languages of different design areas, to understand the causes behind them and their evolution over time. (In other words, I think that "software theory," which this book aims to theorize and put in practice, is not a luxury but a necessity.)

Although the transformations I will be discussing involved many technological and social developments—hardware, software, production practices, and workflows, new job titles and new

Opening titles animation for television series Mad Men. *Imaginary Forces, 2007.*

professional fields—it is appropriate to highlight one particular software application which was at the center of the events. This application is After Effects. In this chapter we will take a close look at its interface, the tools, and its typical use in media design. Introduced in 1993, After Effects was the first software application designed to do animation, compositing, and special effects on the personal computer.[1] Its broad effect on moving image production can be compared to the effects of Photoshop and Illustrator on photography, illustration, and graphic design.

After Effects certainly has its competitors. In the 1990s, companies also widely used more expensive "high-end" software such as Flame, Inferno, or Paintbox that run on specialized graphics workstations, and they are still utilized today. In the 2000s, other programs in the same price category as After Effects such as Apple's Motion, Autodesk's Combustion, and Adobe's Flash have also challenged After Effects' dominance. However, because of its affordability and length of time on the market After Effects continues to be the most popular, most widely used, and best-known application. Consequently, After Effects will be given a privileged role in my account as both the symbol and the key material foundation that made the wide-reaching transformation in moving image culture possible. (When I searched for "best motion graphics software" on the web and checked the answers to this question on a variety of forums, the first program mentioned was always After Effects. As one person put it on one of these forums, "It's pretty much the gold standard. Learn it, love it."[2] In another example, when in 2012 Imaginary Forces—the company most closely associated with the rise of motion graphics in the 1990s—posted descriptions of new jobs in its Los Angeles and NYC offices

---

[1] The NewTek Video Toaster released in 1990 was the first PC-based video production system that included a video switcher, character generation, image manipulation, and animation. Because of their low costs, Video Toaster systems were extremely popular in the 1990s. In the context of this book, After Effects is more important because, as I will explain below, it introduced a new paradigm for moving image design that was different from the familiar video editing paradigm supported by systems such as Toaster and Avid.

[2] John Waskey, http://www.quora.com/What-is-the-best-software-for-creating-motion-graphics (March 4, 2001).

for designers and animators, it listed only one required software application for 2D moving image production: After Effects.[3])

As I will show, After Effects' UI and tools bring together fundamental techniques, working methods, and assumptions of previously separate fields of filmmaking, animation and graphic design. This hybrid production environment, encapsulated in a single software application, is directly reflected in the new visual language it enables—specifically, its focus on exploring aesthetic, narrative, and affective possibilities of hybridization.

The shift to software-based tools in the 1990s affected not only moving image culture but also all other areas of media design. All of them adopted the same type of production workflow. (When the project requires many people and many media elements, the production workflow is called a "pipeline.") In this workflow, designers typically either combine elements created in different software applications, or move the whole project from one application to the next to take advantage of their unique possibilities. And while each design field also employs its own specialized applications (for instance, web designers use Dreamweaver, architects use Revit, and visual effects artists use Nuke and Fusion), they also all use a number of common applications: Photoshop, Illustrator, Final Cut, After Effects, Maya, 3ds Max, and a few others. (If you use open source software like Gimp and CinePaint instead of these commercial applications, your list of key applications will be different, but the workflow would not change.)

The adoption of a production environment and workflow that uses a small number of compatible applications in all areas of creative industries has had many fundamental effects. The professional boundaries between different design fields have become less important. A single designer or a small studio may work on a music video today, a product design tomorrow, an architectural project or a website design the day after, and so on. Another previously fundamental distinction—the scale of a project—also now matters less, and sometimes not at all. Today we can expect to find exactly the same shapes and forms in very small objects (like jewelry), small and medium sized objects (tableware, furniture), large buildings, and even urban designs. (Lifestyle objects, furniture, and

[3] http://www.imaginaryforces.com/jobs/los-angeles/designer/, http://www.imaginaryforces.com/jobs/new-york/2d-animator/ (October 31, 2012).

architectural and urban design by Zaha Hadid's office illustrate this well.[4])

A comprehensive discussion of these and many other effects of software adoption would take more than one book, and therefore in this chapter I only focus on the impact of the software-based workflow on contemporary media design. As we will see, this workflow shapes contemporary design in a number of ways. On the one hand, never before in the history of human visual communication have we witnessed such a variety of visual forms as today. On the other hand, exactly the same techniques, compositions and iconography can now appear in any media. To evoke the metaphor of biological evolution again, we can say that despite seemingly infinite diversity of contemporary media, visual, and spatial "species," they all share some common DNA. Many of the species also share a basic design principle: integration of previously non-compatible techniques of media design—a process which I am going to call "deep remixability." Thus, a consideration of media authoring software and its usage in production would allow us to begin constructing a map of our current media/design universe, seeing how its species are related to each other and revealing the mechanisms behind their evolution.

The adoption of After Effects and related software in the second part of the 1990s quickly led to the adoption of a special term to designate new animated visuals – "motion graphics." Concisely defined in 2003 by Matt Frantz in his Masters thesis as "designed non-narrative, non-figurative based visuals that change over time,"[5] motion graphics include film and television titles, TV graphics, dynamic menus, graphics for mobile media content, and other animated sequences. Typically motion graphics appear as parts of longer pieces: commercials, music videos, training videos, narrative and documentary films, interactive projects. Or at least, this is how it was in 1993; since that time the boundary between motion graphics and everything else has progressively become harder to define. Thus, in the 2008 version of the Wikipedia article about motion graphics, the authors wrote that "the term 'motion graphics' has the potential for less ambiguity than the use of

---

[4] http://www.zaha-hadid.com/

[5] Matt Frantz "Changing Over Time: The Future of Motion Graphics," MFA Thesis, 2003, http://www.mattfrantz.com/thesisandresearch/motiongraphics.html

the term film to describe moving pictures in the 21st century."[6] Certainly, today numerous short moving image works combine live footage, 2D animation, 3D animation, and other techniques equally (as opposed to privileging live action cinematography as many feature films still do), so they all can be called "motion graphics."

Why did I select motion graphics as my central case study of this book, as opposed to any other area of contemporary culture similarly affected by either the switch to a software-based production process, or native to computers? The examples of the former area sometimes called "going digital" are architecture, graphic design, product design, information design, and music; the examples of the latter area are game design, interaction design, user experience design, user interface design, web design, and interactive information visualization. Obviously, most of the new design areas which have "interaction" or "information" as part of their titles—and which emerged since middle of the 1990s have been equally ignored by cultural critics, and therefore—demand as much attention.

My reason has to do with the diversity of new forms—visual, spatial, and temporal—that developed during the rapid growth of the motion graphics field after the introduction of After Effects. If we approach motion graphics in terms of these forms and techniques (rather than only their content), we will realize that they represent a very significant turning point in the history of human communication. Maps, pictograms, hieroglyphs, ideographs, various scripts, alphabet, graphs, projection systems, information graphics, photography, modern language of abstract forms (developed first in European painting in the 1910s and subsequently adopted in graphic design, product design and architecture in the 1920s), the techniques of twentieth-century cinematography, 3D computer graphics, and of course, a variety of "born digital" visual effects—in short practically all communication techniques developed by humans until the 1990s—are now routinely combined in motion graphics projects. Thus, almost all of the previously separate semiotic resources become options within the user's palette (or "toolbox," to use the standard metaphor deployed in media development software). Linguistic, kinetic, spatial, iconic, diagrammatic,

[6] http://en.wikipedia.org/wiki/Motion_graphics

and temporal intelligence can now work together to express what we already knew but could not communicate—as well as generate new messages and experiences whose meanings we have yet to discover.

Although we may still need to figure out how to fully use this new semiotic meta-language, the importance of its emergence is hard to overestimate. In short, the emergence of software-enabled motion graphics is as important historically as the invention of printing, photography, or the Internet.

We will begin by going back to the 1980s. During the heyday of post-modern debates, at least one critic in America noticed the connection between post-modern pastiche and computerization. In his book *After the Great Divide* (1986), Andreas Huyssen wrote, "All modern and avantgardist techniques, forms and images are now stored for instant recall in the computerized memory banks of our culture. But the same memory also stores all of pre-modernist art as well as the genres, codes, and image worlds of popular cultures and modern mass culture." [7] His analysis is accurate—except that these "computerized memory banks" did not really become commonplace for another fifteen years. Only when the Web absorbed enough of the media archives did it become a universal cultural memory bank accessible to all cultural producers. But even for the professionals, the ability to easily integrate multiple media sources within the same project—multiple layers of video, scanned still images, animation, graphics, and typography—only came towards the end of the 1990s.

In 1985, when Huyssen's book was in preparation for publication, I was working for one of the few computer animation companies in the world. The company was located in NYC and it was appropriately called Digital Effects.[8] Each computer animator had his/her own interactive graphics terminal that could show 3D models but only in wireframe and monochrome; to see them fully rendered in color, we had to take turns as the company had only one color raster display which we all shared. The data was stored on bulky magnetic tapes about a foot in diameter; to find the data

---

[7] Andreas Huyssen, "Mapping the Postmodern," in *After the Great Divide* (Bloomington and Indianapolis: Indiana University Press, 1986), p. 196.

[8] Wayne Carlson, *A Critical History of Computer Graphics and Animations. Section 2: The Emergence of Computer Graphics Technology*, http://accad.osu.edu/%7Ewaynec/history/lesson2.html

from an old job was a cumbersome process that involved locating the right tape in the tape library, putting it on a tape drive and then searching for the right part of the tape. We did not have a color scanner, so getting "all modern and avantgardist techniques, forms and images" into the computer was far from trivial. And even if we had had one, there was no way to store, recall, and modify these images. The machine that could do that—Quantel Paintbox—cost over $160,000, which we could not afford. And when in 1986 Quantel introduced Harry, the first commercial non-linear editing system that allowed for digital compositing of multiple layers of video and special effects, its cost similarly made it prohibitive for everybody except network television stations and a few production houses. Harry's capacities were quite limited, because it could record only eighty seconds of broadcast quality video. In the realm of still images, things were not much better: for instance, the digital still store Picturebox released by Quantel in 1990 could hold only 500 broadcast quality images and its cost was similarly very high.

In short, in the middle of the 1980s neither we nor other production companies had anything approaching the "computerized memory banks" imagined by Huyssen. And of course, the same was true for the visual artists that were associated with post-modernism and the ideas of pastiche, collage, and appropriation. In 1986 the BBC produced a documentary *Painting with Light* for which half a dozen well-known painters including Richard Hamilton and David Hockney were invited to work with the Quantel Paintbox. The resulting images were not so different from the normal paintings that these artists were producing without a computer. And while some artists were making reference to "modern and avantgardist techniques, forms and images," these references were painted rather than being directly loaded from "computerized memory banks." Only about ten years later, when relatively inexpensive graphics workstations and personal computers running image editing, animation, compositing, and illustration software became commonplace and affordable for freelance graphic designers, illustrators, and small post-production and animation studios did the situation described by Huyssen start to become a reality.

The results were dramatic. Within the space of less than five years, modern visual culture was fundamentally transformed. Visuals which previously were specific to different media—live action cinematography, graphics, still photography, animation, 3D computer animation, and typography—started to be combined in numerous ways. By the end of the 1990s, the "pure" moving image media became an exception and hybrid media became the norm. However, in contrast to other computer revolutions such as the rise of the World Wide Web around the same time, this revolution was not acknowledged by popular media or cultural critics. What received attention were the developments that affected narrative filmmaking—the use of computer-produced special effects in Hollywood feature films or the inexpensive digital video and editing tools outside of it. But another process which happened on a larger scale—the transformation of the visual language used by all forms of moving images outside of narrative films—has not been critically analyzed. In fact, while the results of these transformations have become fully visible by 1999, at the time of writing I am not aware of a single theoretical article discussing them (at least in English).

One of the reasons is that in this revolution no new media *per se* were created. Just as they did ten years ago, the designers were making still images and moving images. But the aesthetics of these images was now very different. In fact, it was so new that, in retrospect, the post-modern imagery of just ten years ago that at the time looked strikingly different now appears as a barely noticeable blip on the radar of cultural history.

My choice of the starting and ending dates (1993–9) to characterize the development of a new hybrid visual language of moving images is not accidental. Of course, I could have picked different dates—for instance—starting a few years earlier—but since After Effects software which will play the key role in my account was released in 1993, I decided to pick this year as my starting date. And while my ending date also could have been different, I believe that by 1999 the broad changes in the aesthetics of moving images became visible. If you want to quickly see this for yourself, simply compare demo reels from the same visual effects companies made in the early 1990s and the late 1990s (a number of them are available online—look, for instance, at the demo reels of Pacific Data Images, or the demo reels of the Flame system, available for every year starting in 1995.[9]) In work from the

---

[9] http://accad.osu.edu/~waynec/history/lesson6.html;http://area.autodesk.com/flame20#20years.

beginning of the decade, computer imagery in most cases appears by itself—that is, we see whole commercials and promotional videos done in 3D computer animation, and the novelty of this new media is foregrounded. By the end of the 1990s, computer animation becomes just one element integrated in the media mix that also includes live action, typography, and design.

Although these transformations happened only recently, the ubiquity of the new hybrid visual language today is such that it takes an effort to recall how different things looked before. Similarly, the changes in production processes and equipment that made this language possible also quickly fade from public and professional memories. As a way to quickly evoke these changes as seen from the professional perspective, I am going to quote from a 2004 interview with Mindi Lipschultz, who has worked as an editor, producer, and director in Los Angeles since 1979:

> If you wanted to be more creative [in the 1980s], you couldn't just add more software to your system. You had to spend hundreds of thousands of dollars and buy a paintbox. If you wanted to do something graphic—an open to a TV show with a lot of layers—you had to go to an editing house and spend over a thousand dollars an hour to do the exact same thing you do now by buying an inexpensive computer and several software programs. Now with Adobe After Effects and Photoshop, you can do everything in one sweep. You can edit, design, animate. You can do 3D or 2D all on your desktop computer at home or in a small office.[10]

In 1989, the former Soviet satellites of Central and Eastern Europe peacefully liberated themselves from the Soviet Union. In the case of Czechoslovakia, this event came to be referred as the Velvet Revolution—to contrast it to typical revolutions in modern history that were always accompanied by bloodshed. To emphasize the gradual, almost invisible pace of the transformations which occurred in moving image aesthetics between approximately 1993 and 1999, I am going to appropriate the term the Velvet Revolution to refer to these transformations. (Although it may

[10] Mindi Lipschultz, interviewed by *The Compulsive Creative*, May 2004, http://www.compulsivecreative.com/interview.php?intid=12

seem presumptuous to compare political and aesthetics transformations simply because they share the same non-violent quality, it is possible to show that the two revolutions are actually related.)

Finally, before proceeding I should also explain my use of examples. The visual language I am analyzing is all around us today (this may explain why academics have remained blind to it). After globalization, this language is spoken by communication professionals in dozens of countries around the world. You can see for yourself all the examples of various aesthetics I will be mentioning below by simply watching television and paying attention to graphics, going to a club to see a VJ performance, visiting the websites of motion graphics designers and visual effects companies, or opening any book on contemporary design. Nevertheless, below I have included titles of particular projects so the reader can see exactly what I am referring to. (I have chosen works by well-known design studies and artists so you can easily find all of them on the web.) But since my goal is to describe the new cultural language that by now has become practically universal, I want to emphasize that each of these examples can be substituted with numerous others.

The aesthetics of hybridity

In the second half of the 1990s, one of the key identifying features of motion graphics that clearly separated it from other forms of moving image existing until that time, was the central role played by dynamic typography. The term "motion graphics" has been used at least since 1960, when a pioneer of computer filmmaking John Whitney named his new company Motion Graphics. However, until the Velvet Revolution only a handful of people and companies have systematically explored the art of animated typography: Norman McLaren, Saul Bass, Pablo Ferro, R/Greenberg, and a few others.[11] But by the middle of the 1990s moving image sequences or short

---

[11] For a rare discussion of motion graphics prehistory as well as an equally rare attempt to analyze the field by using a set of concepts rather than only presenting the portfolio of individual designers, see Jeff Bellantoni and Matt Woolman, *Type in Motion, 2nd edition* (Thames & Hudson, 2004).

films dominated by moving animated type and abstract graphical elements rather than by live action, started to be produced in large numbers. What was the material cause for motion graphics taking off? After Effects and other related software running on PCs or relatively inexpensive graphics workstations became affordable to smaller design, visual effects, and post-production houses, and soon, to individual designers. Almost overnight, the term "motion graphics" became well known. (As the Wikipedia article about this term points out, "the term 'Motion Graphics' was popularized by Trish and Chris Meyer's book about the use of Adobe After Effects titled 'Creating Motion Graphics.'"[12]) The five-hundred-year-old Gutenberg universe came into motion.

Along with typography, the whole language of twentieth-century graphic design was "imported" into moving image design. While this development did not receive a popular new name of its own, it is obviously at least as important. (Although the term "design cinema" has been used, it never achieved anything comparable to the popularity of "motion graphics.") So while motion graphics were for years limited to film titles and therefore only used typography, today the term "motion graphics" is often used to refer to moving image sequences that combine moving type and design elements. But we should recall that while in the twentieth century typography was often used in combination with other design elements, for 500 years it commanded its own word. Therefore I think it is important to consider the two kinds of "import" operations that took place during the Velvet Revolution—typography and twentieth century graphic design—as two distinct historical developments.

While motion graphics definitely exemplify the changes that took place during the Velvet Revolution, these changes are broader. Simply put, the result of the Velvet Revolution is *a new hybrid visual language of moving images in general*. This language is not confined to particular media forms. And while today it manifests itself most clearly in non-narrative forms, it is also often present in narrative and figurative sequences and films.

Here are a few examples. A music video may use live action while also employing typography and a variety of transitions done

[12] http://en.wikipedia.org/wiki/Motion_graphic

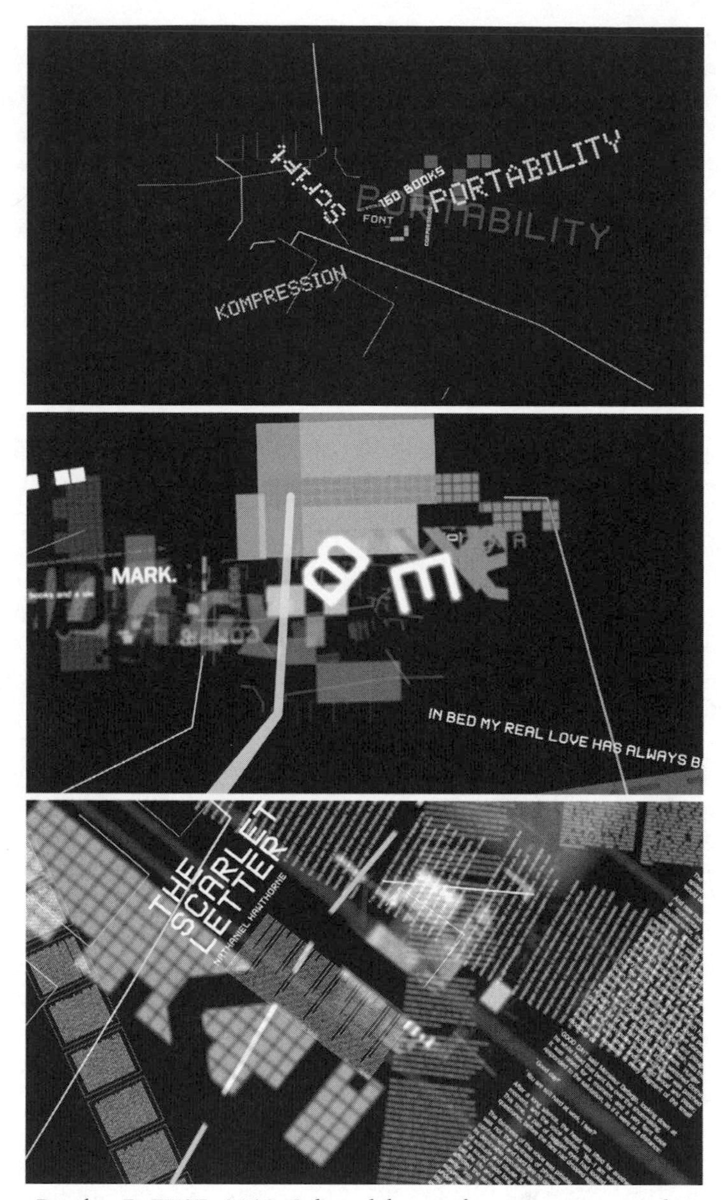

Sony Reader. *D-FUSE, 2009. Selected frames from a motion graphics video.*

with computer graphics (video for "Go" by Common, directed by Convert/MK12/Kanye West, 2005). Another music video may embed the singer within an animated painterly space (video for Sheryl Crow's "Good Is Good," directed by Psyop, 2005). A commercial may superimpose charts, data displays, and data visualizations on top of live action (TV ad for Thomson Reuters by MK12, 2012). The title sequence may contrast 2D flat figures and deep 3D perspectival space (titles for *Mad Men* by Imaginary Forces, 2007). A short film may mix typography, stylized 3D graphics, animated 2D design elements, and live action (*Itsu* for Plaid, directed by the Pleix collective, 2002). (Sometimes, as I have already mentioned, the term "design cinema" is used to differentiate such short independent films organized around design, typography, and computer animation rather than live action from similar "motion graphics" works produced for commercial clients.)

In some cases, the juxtaposition of different media is clearly visible (video for "Don't Panic" by Coldplay, 2001; titles for the television show *The Inside*, 2005; commercial "Nike – Dynamic Feet," 2005, all by Imaginary Forces). In other cases, a sequence may move between different media so quickly that the shifts are barely noticeable (GMC Denali "Holes" commercial by Imaginary Forces, 2005). Yet in other cases, a commercial or movie title may feature continuous action shot on video or film, with the image periodically changing from a more natural to a highly stylized look.

Such media hybridity does not necessarily manifest itself in a collage-like aesthetics that foregrounds the juxtaposition of different media and different media techniques. As a very different example of what media hybridity can result in, consider a more subtle aesthetics well captured by the name of the software that to a large extent made the hybrid visual language possible: After Effects. This name anticipated the changes in visual effects that took place a number of years later. In the 1990s computers were used to create highly spectacular special effects or "invisible effects,"[13] but towards the end of that decade we see something

---

[13] *Invisible effect* is the standard industry term. For instance, the film *Contact*, directed by Robert Zemeckis, was nominated for the 1997 VFX HQ Awards in the following categories: Best Visual Effects, Best Sequence (The Ride), Best Shot (Powers of Ten), Best Invisible Effects (Dish Restoration), and Best Compositing. http://www.vfxhq.com/1997/contact.html

else emerging: a new visual aesthetics that goes "beyond effects." In this aesthetics, the whole project—whether a music video, a TV commercial, a short film, or a large segment of a feature film—has a special look in which the enhancement of live-action material is not completely invisible but at the same time does not call attention to itself the way special effects tended to do in the 1990s (examples: Reebok I-Pump "Basketball Black" commercial and *The Legend of Zorro* main title, both by Imaginary Forces, 2005; "Fage 'Plain'" commercial by Psyop, 2011).

Although the particular aesthetic solutions vary from one video to the next and from one designer to another, they share the same logic: the simultaneous appearance of multiple media within the same frame. Whether these media are openly juxtaposed or almost seamlessly blended together is less important than the fact of this copresence itself. (Again, note that each of the examples above can be substituted with numerous others.)

Hybrid visual language is also now common to a large proportion of short "experimental" and "independent" (i.e., not commissioned by commercial clients) videos being produced for media festivals, the web, mobile media devices, and other distribution platforms.[14] Many visuals created by VJs and "live cinema" artists are also hybrid, combining video, layers of 2D imagery, animation, and abstract imagery generated in real time.[15] And as the animations of artists Jeremy Blake, Ann Lislegaard, and Takeshi Murata that I will discuss below demonstrate, at least some of the works created explicitly for art-world distribution similarly choose to use the same language of hybridity.

---

[14] In December 2005, I attended the Impact media festival in Utrecht and asked the festival director what percentage of the submissions they received that year featured hybrid visual language as opposed to "straight" video or film. His estimate was about 50 percent. In January 2006, I was part of the review team that judged the projects of students graduating from SCI-ARC, a well-known research-oriented architecture school in Los Angeles. According to my informal estimate, approximately half the projects featured complex curved geometry made possible by Maya, a modeling software now commonly used by architects. Given that both After Effects and Maya's predecessor, Alias, were introduced in the same year—1993—I find this quantitative similarity in the percentage of projects that use new languages made possible by this software quite telling.

[15] For examples, consult Paul Spinrad, ed., *The VJ Book: Inspirations and Practical Advice for Live Visuals Performance* (Feral House, 2005).

Today, narrative features rarely mix different graphical styles within the same frame. However, gradually growing number of films do feature highly stylized aesthetics that would have previously been identified with illustration rather than filmmaking. The examples are the Wachowski's *Matrix* series (1999–2003), *Immortal* by Enki Bilal (2004), Robert Rodriguez' *Sin City* (2005) and *The Spirit* (2008), Zack Snyder's *300* (2007) and *Watchmen* (2009), James Cameron's *Avatar* (2009), Tim Burton's *Alice In Wonderland* (2010) and Martin Scorsese's *Hugo* (2011). These feature films are examples of now fully established practice to shoot a large portion of a feature film using a "digital backlot" (i.e., a green screen).[16] Consequently, most or all shots in such films are created by composing the footage of actors with computer-generated sets and other visuals.

These films do not juxtapose their different media as dramatically as motion graphics. Nor do they strive for the seamless integration of CGI (computer-generated imagery) visuals and live action that characterized the earlier special-effects features of the 1990s, such as *Terminator 2* (1991) and *Titanic* (1997) by James Cameron. Instead, they explore the space in between juxtaposition and complete integration.

Matrix, *Sin City*, *300*, *Alice In Wonderland*, *Hugo*, and other films shot on a digital backlot combine multiple media to create a new stylized aesthetics that cannot be reduced to the already familiar look of live-action cinematography or 3D computer animation. Such films display exactly the same logic as short motion graphics works, which at first sight might appear to be very different. This logic is also the same as that which we observe in the creation of new hybrids in biology. That is, the result of the hybridization process is not simply a mechanical sum of the previously existing parts but a new "species"—a new kind of visual aesthetics that did not exist previously.

In TV commercials produced in the 2000s, this highly stylized visual aesthetics became one of the key looks of the decade. Many layers of live footage, 3D and 2D animated elements, particle effects, and other media elements are blended to create a seamless whole. This result has the crucial codes of realism (perspective

---

[16] http://en.wikipedia.org/wiki/Digital_backlot

foreshortening, atmospheric perspective, correct combination of lights and shadows), but at the same time enhances visible reality. (I cannot call this aesthetics "hyperreal" since the hybrid images assembled from many layers and types of media look quite different from the works of the hyperrealist artists such as Denis Peterson that visually look like standard color photographs.) Strong gray-scale contrast, high color saturation, tiny waves of particles emulating from moving objects, extreme close-ups of textured surfaces (water drops, food products, human skin, finishes of consumer electronics devices, etc.), the contrasts between the natural uneven textured surfaces and smooth 3D renderings and 2D gradients, the rapidly changing composition and camera position and direction, and other devices heighten our perception. (For examples of all these strategies, you can, for example, look at the commercials made by Psyop.[17])

We can say that these commercials create a "map" which is bigger than the territory being mapped, because they show more details and texture spatially, and at the same time compress time, moving through information more rapidly. We can also make a comparison with the Earth observation satellites which circle the planet, capturing its whole surface in detail impossible for any human observer to see—just as a human being cannot simultaneously see the extreme close-up of the surfaces and details of the movements of objects presented in the fictional space of a commercial.

In summary, the result of the shift to a software production environment in moving image creation is a new visual language. Just as with the individual software techniques that make it possible, this language as a whole inherits the traits of previous image media—filmmaking, cel and puppet animation, computer animation, photography, painting, graphic design and typography. However, it is not reducible to any of these media. Rather, it is a true hybrid—the offspring of twentieth-century image mediums that shares common traits with all of them but has its own distinct identity.

Let us now look at two short films in detail to see how aesthetics in hybridity works across a whole film. Blake's *Sodium Fox* (2005) and Murata's *Untitled (Pink Dot)* (2007) offer excellent examples

[17] http://www.psyop.tv/projects/live-action/ (October 31, 2012)

of the new hybrid visual language that currently dominates moving-image culture. Among many well-known artists working with digital moving images, Blake was the earliest and most successful in developing his own style of hybrid media. His video *Sodium Fox* is a sophisticated blend of drawings, paintings, 2D animation, photography, and effects available in software. Using a strategy commonly employed by artists in relation to commercial media in the twentieth century, Blake slows down the fast-paced rhythm of motion graphics as they are usually practiced today. However, despite the seemingly slow pace of his film, it is as informationally dense as the most frantically changing motion graphics used in clubs, music videos, television station IDs, and so on. *Sodium Fox* creates this density by exploring in an original way the basic feature of the software-based production environment in general and programs such as After Effects in particular—namely, the construction of an image from potentially numerous layers. Of course, traditional cel animation as practiced in the twentieth century also involved building up an image from a number of superimposed transparent cels, with each one containing some of the elements that together make up the whole image. For instance, one cel could contain a face, another lips, a third hair, yet another a car, and so on.

With computer software, however, designers can precisely control the transparency of each layer; they can also add different visual effects, such as blur, between layers. As a result, rather than creating a visual narrative based on the motion of a few visual elements through space (as was common in twentieth-century animation, both commercial and experimental), designers now have many new ways to create animation. Exploring these possibilities, Blake crafts his own visual language in which visual elements positioned on different layers are continuously and gradually "written over" each other. If we connect this new language to twentieth-century cinema rather than to cel animation, we can say that rather than fading in a new frame as a whole, Blake continuously fades in separate parts of an image. The result is an aesthetics that balances visual continuity with a constant rhythm of visual rewriting, erasing, and gradual superimposition.

Like *Sodium Fox*, Murata's *Untitled (Pink Dot)* also develops its own language within the general paradigm of media hybridity. Murata creates a pulsating and breathing image that has a distinctly biological feel to it. In the last decade, many designers and artists

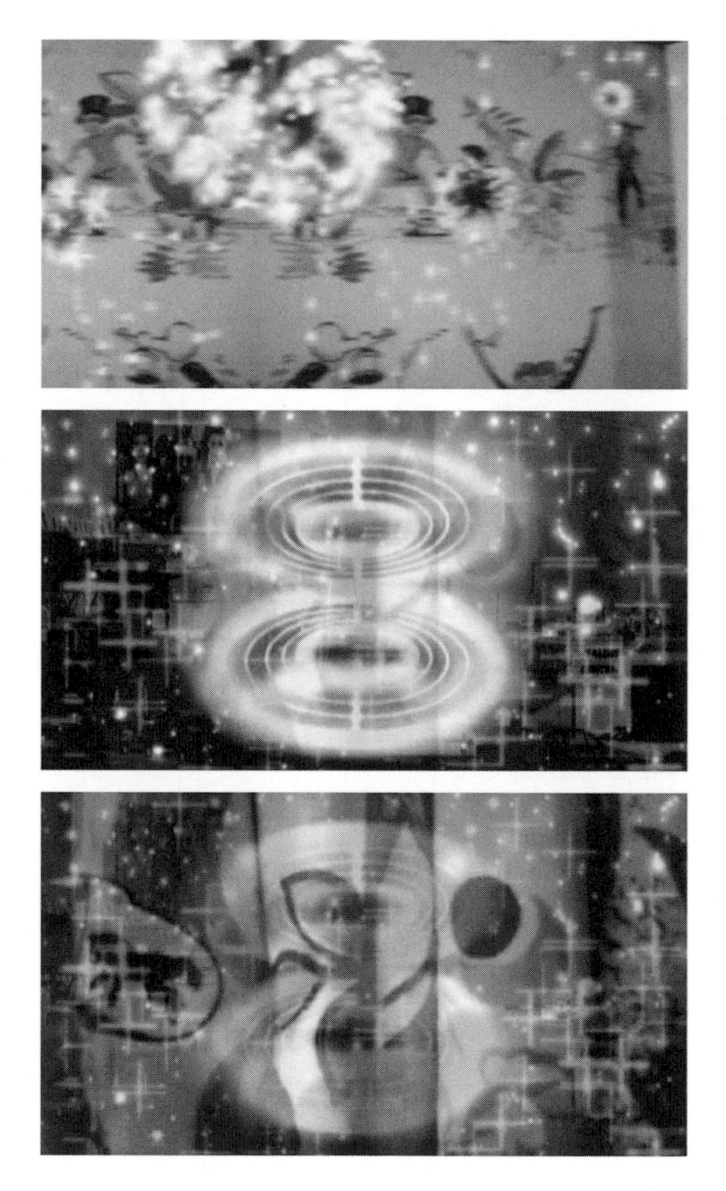

Sodium Fox. *Jeremy Blake, 2005. Selected frames from a 14-minute digital video.*

Untitled (Pink Dot). *Takeshi Murata, 2007. Selected frames from a five-minute digital video.*

have used biologically inspired algorithms and techniques to create animal-like movements in their generative animations and interactives. However, in the case of *Untitled (Pink Dot)*, the image as a whole seems to come to life.

To create this pulsating, breathing-like rhythm, Murata transforms live-action footage (scenes from 1982 movie *Rambo: First Blood*) into a flow of abstract color patches (sometimes they look like oversize pixels, and at other times they may be taken for artifacts of heavy image compression). But this transformation never settles into a final state. Instead, Murata constantly adjusts its degree. (In terms of the interfaces of media software, this would correspond to animating a setting of a filter or an effect.) One moment we see almost unprocessed live imagery; the next moment it becomes a completely abstract pattern; in the following moment, parts of the live action image again become visible, and so on.

In *Untitled (Pink Dot)* the general condition of media hybridity is realized as a permanent metamorphosis. True, we still see some echoes of movement through space, which was the core method of pre-digital animation. (Here this is the movement of the figures in the shots from *Rambo*.) But now the real change that matters is the one between different media aesthetics: between the texture of a film and the pulsating abstract patterns of flowing patches of color, between the original "liveness" of human figures in action as captured on film and the highly exaggerated artificial liveness they generate when processed by a machine.

Visually, *Untitled (Pink Dot)* and *Sodium Fox* do not have much in common. However, both films share the same strategy: creating a visual narrative through continuous transformations of image layers, as opposed to discrete movements of graphical marks or characters, common to both the classic commercial animation of Disney and the experimental classics of Norman McLaren, Oskar Fischinger, and others. Although we can assume that neither Blake nor Murata has aimed to achieve this consciously, in different ways each artist stages for us the key technical and conceptual change that defines the new era of media hybridity. Media software allows the designer to combine any number of visual elements regardless of their original media and to control each element in the process. This basic ability can be explored in numerous visual aesthetics. The films of Blake and Murata, with their different temporal rhythms and different logics of media combination, exemplify this diversity. Blake layers various still graphics, text, animation, and effects, dissolving elements in and out. Murata processes live footage to create a constant image flow in which the two layers—live footage and its processed result—seem to constantly push each other out.

Deep remixability

I believe that "media hybridity" constitutes a new fundamental stage in the history of media. It manifests itself in different areas of software culture and not only in moving images—although the latter does offer a particularly striking example of this new cultural logic at work. Here the media authoring software environment became a kind of Petri dish where the techniques and tools of computer animation, live cinematography, graphic design, 2D animation, typography, painting and drawing can interact, generating new hybrids. And as the examples above demonstrate, the results of this process of hybridity are a new aesthetics and new "media species" that cannot be reduced to the sum of the media that went into their creation.

Can we understand the new hybrid language of moving image as a type of *remix*? From its beginnings in music culture in the 1980s, during the 1990s remix has gradually emerged as the dominant aesthetics of the era of globalization, affecting and re-shaping everything from music and cinema to food and fashion. (If Fredric Jameson once referred to post-modernism as "the cultural logic of late capitalism," we can perhaps call remix "the cultural logic of networked global capitalism.") A number of authors have already traced remix effects in many cultural areas, ranging from children's use of media in Japan (Mimi Ito) to web culture (Eduardo Navas). The representative books include *Rhythm Science* (D. J. Spooky, 2004), *Remix: Making Art and Commerce Thrive in the Hybrid Economy* (Lawrence Lessig, 2008), *Mashed Up: Music, Technology, The Rise of Configurable Culture* (Aram Sinnreich, 2010), *Mashup Cultures* (edited by Stefan Sonvilla-Weiss, 2010) and *Remix Theory: The Aesthetics of Sampling* (Eduardo Navas, 2012).[18]

[18] Mizuko Ito, "Mobilizing the Imagination in Everyday Play: The Case of Japanese Media Mixes," in *International Handbook of Children, Media, and Culture,*

I believe that the combinatory mechanisms responsible for the evolution of the "computer metamedium" in general, and the new hybrid visual aesthetics emerging in the 1990s can indeed be considered as a type of remix—if we make one crucial distinction. Typical remix combines *content* within the same media or content from different media. For instance, a music remix may combine music elements from any number of artists; anime music videos may combine parts of anime films and music taken from a music video. Professionally produced motion graphics and other moving-image projects also routinely mix together content in the same media and/or from different media. For example, in the beginning of the "Go" music video (Convert/MK12/Kanye West, 2005), the video rapidly switches between live-action footage of a room and a 3D model of the same room. Later, the live-action shots also incorporate a computer-generated plant and a still photographic image of mountain landscape. Shots of a female dancer are combined with elaborate animated typography. The human characters are transformed into abstract animated patterns. And so on.

Such remixes of content from different media are definitely common today in moving-image culture. In fact, I began discussing the new visual language by pointing out that in the case of short forms they now constitute the rule rather than the exception. But this type of remix is only one aspect of the "hybrid revolution." For me its essence lies in something else. Let us call it "deep remixability." *Today designers remix not only content from different media but also their fundamental techniques, working methods, and ways of representation and expression.* United within the common software environment the languages of cinematography, animation, computer animation, special effects, graphic design, typography, drawing, and painting have come to form a new *metalanguage*. A work produced in this new metalanguage can

---

Sonia Livingstone and Kirsten Drotner, (eds) (Sage Publications, 2008); Paul D. Miller, *Rhythm Science* (MIT Press, 2004); Lawrence Lessig. *Remix: Making Art and Commerce Thrive in the Hybrid Economy* (Penguin Press HC, 2008); Aram Sinnreich, *Mashed Up: Music, Technology, and the Rise of Configurable Culture* (University of Massachusetts Press, 2010); Stefan Sonvilla-Weiss, *Mashup Cultures* (Springer, 2010); Eduardo Navas, *Remix Theory: The Aesthetics of Sampling* (Springer Vienna Architecture, 2012).

use all the techniques, or any subset of these techniques, that were previously unique to these different media.

We may think of this new metalanguage of moving images as a large library of all previously known techniques for creating and modifying moving images. A designer of moving images selects techniques from this library and combines them in a single sequence or a single frame. But this clear picture is deceptive. How exactly does s/he combine these techniques? When you remix content, it is easy to imagine different texts, audio samples, visual elements, or data streams positioned side by side. Imagine a typical twentieth-century collage except that it now moves and changes over time. But how do you remix the techniques themselves?

In the cases of hybrid media interfaces that we have already analyzed (such as Acrobat's interface), "remix" of techniques means simple combination. Different techniques literally appear next to each in the application's UI. Thus, in Acrobat, a forward and backward button, a zoom button, a "find" tool, and others are positioned one after another on a toolbar above the open document. Other techniques appear as tools listed in vertical pull-down menus: spell, search, email, print, and so on. We find the same principles in the interfaces of all media authoring and access applications. The techniques borrowed from various media and the new born-digital techniques are presented side-by-side using tool-bars, pull-down menus, toolboxes and other UI design conventions.

Such an "addition of techniques" that exist in a single space side by side without any deep interactions are also indirectly present in remixes of content well familiar to us, be it fashion designs, architecture, collages, or motion graphics. Consider a hypothetical example of a visual design that combines drawn elements, photos, and 3D computer graphics forms. Each of these visual elements is the result of the use of particular media techniques of drawing, photography, and computer graphics. Thus, while we may refer to such cultural objects as remixes of content, we are also justified in thinking about them as remixes of techniques. This applies equally well to pre-digital design, when a designer would use separate physical tools or machines, and to contemporary software-driven design, where s/he has access to all these tools in a few compatible software applications.

As long as the pieces of content, interface buttons, or techniques are simply added rather than integrated together, we do not need a

00–07

00–11

00–16

Music video for Go! *By Common. Kanye West, MK12 and Convert, 2005. Selected frames from a four-minute video. A number below each frame indicates time code (in seconds) for this frame.*

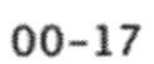

00-17

00-21

00-23

special term such as "deep remix." This, for me, is still "remix" the way this term is commonly used. But in the case of moving image aesthetics we also encounter something else. Rather than a simple addition, we also find interactions between previously separate techniques of cel animation, cinematography, 3D animation, design, and so on—interactions which were unthinkable before. (The same argument can be made in relation to other types of cultural objects and experiences created with media authoring software such as visual designs and music.)

I believe that this is something that neither pioneers of computer media of the 1960s–1970s nor the designers of the first media authoring applications that started to appear in the 1980s intended. However, once all these media techniques met within the same software environment—and this was gradually accomplished throughout the 1990s—they started interacting in ways that could never have been predicted or even imagined previously.

For instance, while particular media techniques continue to be used in relation to their original media, they can also be applied to other media. Photoshop filters, which we previously analyzed, illustrate well this "cross-over" effect: techniques that were originally part of a particular media type can now be applied to other media types. For instance, neon glow, stained glass, lens flare, etc. can be applied to photographs and sketches. (More precisely, they can be applied to whatever is currently loaded in the graphics memory and appears in the image window—a set of pixels that carry the results of all previously applied filters and other manipulations.)

Here are typical examples of the crossover strategy as it is used in moving image design. Type is choreographed to move in 3D space; motion blur is applied to 3D computer graphics; algorithmically generated fields of particles are blended with live-action footage to give them an enhanced look; a virtual camera is made to move around a virtual space filled with 2D drawings. In each of these examples, the technique that was originally associated with a particular medium—cinema, cel animation, photorealistic computer graphics, typography, graphic design—is now applied to a different media type. Today a typical short film or a sequence may combine many of such pairings within the same frame. The result is a hybrid, intricate, complex, and rich media language—or rather, numerous languages that share the logic of deep remixability.

In fact, such interactions among virtualized media techniques define the aesthetics of contemporary moving image culture. This is why I have decided to introduce a special term—*deep remixability*. I wanted to differentiate more complex forms of interactions between techniques (such as cross over) from the simple remix (i.e. addition) of media content and media techniques with which we are all familiar, be it music remixes, anime video remixes, 1980s postmodern art and architecture, and so on.

For concrete examples of the "crossover effect" which exemplifies deep remixability, we can return to the same "Go" video and look at it again, but now from a new perspective. Previously I have pointed out the ways in which this video—typical of the short format moving images works today—combines visual elements of *different media types*: live action video, still photographs, procedurally generated elements, typography, etc. However, exactly the same shots also contain rich examples of the *interactions between techniques*, which are only possible in a software-driven design environment.

As the video begins, a structure made up from perpendicular monochrome blocks and panels simultaneously grows rapidly in space and rotates to settle into a position which allows us to recognize it as a room (00:07–00:11). As this move is being completed, the room is transformed from an abstract geometric structure into a photo-realistically rendered one: furniture pops in, wood texture rolls over the floor plane, and a photograph of a mountain view fills a window. Although such different styles of CG rendering have been available in animation software since the 1980s, a particular way in which this video opens with a visually striking abstract monochrome 3D structure is a clear example of deep remixability. When in the middle of the 1990s graphic designers started to use computer animation software, they brought their training, techniques, and sensibilities to computer animation that until that time was used in the service of photorealism. The strong diagonal compositions, the deliberately flat rendering, and the choice of colors in the opening of the "Go" video subordinates CG photorealistic techniques to a visual discipline specific to modern graphic design. The animated 3D structure references the Suprematism of Malevich and Lissitzky, which played a key role in shaping the grammar of modern design—and which, in our example, has become a conceptual "filter" that has transformed the CG field.

After a momentary stop to let us take in the room, which is now largely completed, a camera suddenly rotates 90 degrees (00:15 – 00:17). This physically impossible camera move is another example of deep remixability. While animation software implements the standard grammar of twentieth-century cinematography—a pan, a zoom, a dolly, etc.—the software, of course, does not have the limitations of a physical world. Consequently a camera can move in an arbitrary direction, follow any imaginable curve and do this at any speed. Such impossible camera moves become standard tools of contemporary media design and twenty-first-century cinematography, appearing with increased frequency in feature films since the middle of the 2000s. Just as Photoshop filters which can be applied to any visual composition, virtual camera moves can also be superimposed, so to speak, on any visual scene regardless of whether it was constructed in 3D, procedurally generated, captured on video, photographed, or drawn—or, as in the example of the room from "Go" video, is a combination of these different media.

Playing the video forward (00:15–00:22), we notice yet another previously impossible interaction between media techniques. The interaction in question is a lens reflection, which is slowly moving across the whole scene. Originally an artifact of a camera technology, lens reflection was turned into a filter—i.e., a technique which can now be "drawn" over any image constructed with all other techniques available to a designer. (This important type of software technique, one which originated as artifacts of physical or electronic media technologies, will be discussed in more details in the concluding section of this chapter.) If you wanted more proof that we are dealing here with a visual technique, note that this "lens reflection" is moving while the camera remains perfectly still (00:17–00:22)—a logical impossibility, which is sacrificed in favor of a more dynamic visual experience.

I referred to the new language of moving imagery as a "meta-language." Since our discussion has so far relied on the term "computer metamedium," I should explain the connection between these two terms.

The acceleration of the speed of social, technological and cultural change in the second part of the twentieth century has led to the frequent use of 'meta-,' 'hyper-,' and 'super-' in cultural theory and criticism. From Superstudio (a conceptual architectural group active in the 1960s), Ted Nelson's Hypermedia and

Alan Kay's metamedium to the more recent Supermodernism and Hypermodernity,[19] these terms may be read as attempts to capture the feeling that we have passed a point of singularity and are now moving at warp speed. Like the cosmonauts of the 1960s observing the Earth from the orbits of their spaceships and seeing it for the first time as a single object, we are looking down at human history from a new higher orbit. This connotation seems to fit Alan Kay's conceptual and practical redefinition of a digital computer as a metamedium that contains most existing medium technologies and techniques and also allows invention of many new ones.

While the term "metalanguage" has precise meanings in logic, linguistics, and computing, here I am using it in a sense similar to Alan Kay's use of "meta" in "computer metamedium." Normally a "metalanguage" refers to a separate formal system for describing mediums or cultural languages—the way grammar describes how a particular natural language works. But this not how Kay uses "meta" in "metamedium." As he uses it, it stands for gathering/including/collecting—in short, bringing previously separate things together.

Let us imagine this computer metamedium as a large and continuously expanding set of resources. It includes all media creation/manipulation techniques, interaction techniques and data formats available to programmers and designers in the current historical moment. Everything from sort and search algorithms and pull-down menus to hair and water rendering techniques, video games AI, and multi-touch interface methods—it is all there.

If we look at how these resources are used in different cultural areas to create particular kinds of content and experiences, we will see that each of them only uses a subset of these resources. For example, the graphical interfaces of today's popular computer operating systems (Windows, Linux, Mac OS) use static icons. In contrast, in some consumer electronics interfaces (such as certain mobile phones) all icons are animated loops.

Moreover, the use of a subset of all existing elements is not random but follows particular conventions. Some elements always go together. In other cases, the use of one element means that we

[19] http://en.wikipedia.org/wiki/Hypermodernity

are unlikely to find some other element. In other words, different forms of digital media use different subsets from a complete set of techniques contained in a computer metamedium—and this use follows distinct patterns.

If you notice the parallels with what cultural critics usually call an "artistic language," a "style," or a "genre," you are right. Any single work of literature or works of a particular author or literary movement uses only some of the existing literary techniques, and this use follows some patterns. The same goes for cinema, music and all other recognized cultural forms. This allows us to talk about a style of a particular novel or film, a style of an author as a whole, or a style of a whole artistic school. (Film scholars David Bordwell and Kristin Thompson call this a "stylistic system" which they define as a "patterned and significant use of techniques." For cinema, they divide these techniques into four categories: mise-en-scène, cinematography, editing, and sound.[20]) When a whole cultural field can be divided into a small number of distinct groups of works with each group sharing some patterns, we usually talk about "genres." For instance, theoreticians of Ancient Greek theatre distinguished between comedies and tragedies and prescribed the rules each genre should follow, while today companies use automatic software to classify blogs into different genres.

If by medium we mean a set of standard technological resources, be it a physical stage or a film camera, lights and film stock, we can see that each medium usually supports multiple artistic languages/styles/genres. For example, a medium of twentieth-century filmmaking supported Russian Montage of the 1920s, Italian Neorealism of the 1940s, French New Wave of the 1960s, Hong Kong fantasy Kung Fu films of the 1980s, Chinese "fifth-generation" films of the 1980s–1990s, etc.

Similarly, a computer metamedium can support multiple cultural or artistic metalanguages. In other words, in the theoretical scheme I am proposing, there is only one metamedium—but many metalanguages.

So what is a metalanguage? If we define an artistic language as a patterned use of a selected number of a subset of the techniques

[20] David Bordwell and Kristin Thompson, *Film Art: an Introduction*, 5th edition (The McGraw-Hill Companies, 1997), p. 355.

available in a given medium,[21] *a metalanguage is a patterned use of a subset of all the techniques available in a computer metamedium*. But not just any subset. It only makes sense to talk about a metalanguage (as opposed to a language) if the techniques it uses come from previously distinct cultural languages. As an example, consider a metalanguage of popular commercial virtual globes (Google Earth, Microsoft Bing Maps). These applications systematically combine different types of media formats and media navigation techniques that previously were separate. These combinations follow common patterns. Another example will be a metalanguage common to many graphical interface users (recall my analysis of Acrobat's interface, which combines metaphors drawn from different media traditions).

Since moving images today systematically combine techniques of different visual media that almost never met until middle of the 1990s, we are justified in using the term "metalanguage" in their case. Visual design today has its own metalanguage, itself is a subset of the metalanguage of moving images. The reason is that a designer of moving images has access to all the techniques of a visual designer plus extra techniques since s/he is working with additional dimension of time. These two metalanguages also largely overlap in patterns that are common to them—but there are some important differences. For instance, today's moving image works often feature a continuous movement through a 3D space that may contain various 2D elements. In contrast, visual designs for print, web, products, or other applications are usually two-dimensional—they assemble elements over an imaginary flat surface. (I think that the main reason for this insistence on flatness is that these designs often exist next to large blocks of text that already exist in 2D.)

Layers, transparency, compositing

So far I have focused on describing the aesthetics of moving images that emerged from the Velvet Revolution. While continuing this investigation, we will now pay more attention to an analysis of the new software production environment that made this aesthetics

[21] This definition is adopted from Bordwell and Thompson, *Film Art*, p. 355.

possible. The following sections of this chapter will look at the tools offered by After Effects and other media authoring applications, their user interfaces, and the ways these applications are used together in production (i.e. design workflow). Rather than discussing all the tools and interface features, I will highlight a number of fundamental assumptions behind them—ways of understanding what a moving image project is—which, as we will see, are quite different from how it was understood during the twentieth century.

Probably most dramatic among the changes that took place during 1993–99 was the new ability to combine multiple levels of imagery with varying degrees of transparency via digital compositing. If you compare a typical music video or a TV advertising spot *circa* 1986 with its counterpart *circa* 1996, the differences are striking. (The same holds for other areas of visual design.) As I have already noted, in 1986 "computerized memory banks" were very limited in their storage capacity and prohibitively expensive, and therefore designers could not quickly and easily cut and paste multiple image sources. But even when they could assemble multiple visual references, a designer only could place them next to, or on top of each other. S/he could not modulate these juxtapositions by precisely adjusting transparency levels of different images. Instead, s/he had to resort to the same photo-collage techniques popularized in the 1920s. In other words, the lack of transparency restricted the number of different image sources that could be integrated within a single composition without it starting to look like certain photomontages or photo-collages of John Heartfield, Hannah Hoch, or Robert Rauschenberg—a mosaic of fragments without any strong dominant.[22]

In addition to allowing the superimposition of many transparent layers, digital compositing also made trivial another operation that was previously very cumbersome. Until the 1990s, different media types such as hand-drawn animation, lens-based recordings, and typography practically never appeared within the same frame. Instead, animated commercials, publicity shorts, industrial films, and some feature and experimental films that did include multiple media

[22] In the case of video, one of the things that made combining multiple visuals difficult was the rapid degradation of the video signal when an analog video tape was copied more than a couple of times. Such a copy would no longer meet broadcasting standards.

usually placed them in separate shots. A few directors have managed to build whole aesthetic systems out of such temporal juxtapositions—most notably, Jean-Luc Godard. In his 1960s films such as *Weekend* (1967) Godard cut bold typographic compositions in between live action creating what can be called a "media montage" (as opposed to a montage of live action shots, as developed by the Russians in the 1920s). Also in the 1960s, pioneering motion graphics designer Pablo Ferro, who appropriately called his company Frame Imagery, created promotional shorts and TV graphics that played on juxtapositions of different media replacing each other in rapid succession.[23] In a number of Ferro's spots, static images of different letterforms, line drawings, original hand-painted artwork, photographs, very short clips from newsreels, and other visuals come one after another with machine gun speed.

Within cinema, the superimposition of different media within the same frame was usually limited to the two media placed on top of each other in a standardized manner—i.e. static letters appearing on top of still or moving lens-based images in feature film titles. In the 1960s, both Ferro and another motion graphics pioneer Saul Bass created a few remarkable title sequences in which visual elements of different origin were systematically overlaid together more dynamically—such as the opening for Hitchcock's *Vertigo* designed by Bass (1958). (Bass's 1959 title sequence for *North by Northwest* is considered to be the first to use type in motion).

But I think it is fair to say that such complex juxtapositions of media within the same frame were rare exceptions in the otherwise "unimedia" universe, where filmed images appeared in feature films and hand-drawn images appeared in animated films. The only twentieth-century feature film director I know of who has built his unique aesthetics by systematically combining different media within the same frame was Karel Zeman. Thus, a typical shot by Zeman may contain filmed human figures, an old engraving used for background, and a miniature model.[24]

The achievements of these directors and designers are particularly remarkable given the difficulty of combining different media

---

[23] Jeff Bellantoni and Matt Woolman, *Type in Motion* (Rizzoli, 1999), pp. 22–9.

[24] While special effects in feature films often combined different media, they were used together to create a single illusionistic space, rather than juxtaposed for the aesthetic effect such as in films and titles by Godard, Zeman, Ferro, and Bass.

within the same frame during the film era. To do this required utilizing the services of special effects departments or separate companies which used optical printers. Techniques that were cheap and more accessible, such as double exposure, were limited in their precision. So while a designer of static images could at least cut and paste multiple elements within the same composition to create a photomontage, to produce the equivalent effect with moving images was far from trivial.

To put this in more general terms, we can say that before the computerization of the 1990s, the designer's capacities to access, manipulate, remix, and filter visual information, whether still or moving, were quite restricted. In fact, they were practically the same as a hundred years earlier—regardless of whether filmmakers and designers used in-camera effects, optical printing, or video keying. In retrospect, we can see they were at odds with the flexibility, speed, and precision of data manipulation that was already available to most other by then computerized professional fields—sciences, engineering, accounting, management, etc. Therefore it was only a matter of time before all image media would be turned into digital data and illustrators, graphic designers, animators, film editors, video editors, and motion graphics designers would start to manipulate them via software instead of their traditional tools. But this is only obvious today—after the Velvet Revolution has taken place.

In 1985 Jeff Stein directed a music video for the single "You Might Think" by new wave band *The Cars*. This video was one of the first to systematically use computer graphics; it had a big impact in the design world, and MTV gave it the first prize in its first annual music awards. Stein managed to create a surreal world in which a video cutout of the singing head of the band member was animated over different video backgrounds. In other words, Stein took the aesthetics of animated cartoons—2D animated characters superimposed over a 2D background—and recreated it using video imagery. In addition, simple computer animated elements were also added in some shots to enhance the surreal effect. This was shocking because no one had ever seen such juxtapositions before. Suddenly, modernist photomontage came alive. But ten years later, such moving video collages became not only commonplace—they also became more complex, more layered, and more subtle. Instead of two or three, a composition could now feature hundreds and even thousands of layers. And each layer could have its own level of transparency.

In short, *digital compositing* now allowed the designers of moving images to easily *combine any number of visual elements regardless of the media in which they originated and to control each element in the process*. I can make an analogy between multitrack audio recording and digital compositing. In multitrack recording, each sound track can be manipulated individually to produce the desired result. Similarly, in digital compositing each visual element can be independently modulated in a variety of ways: resized, recolored, animated, etc. Just as the music artist can focus on a particular track while muting all other tracks, a designer often turns of all visual tracks except the one s/he is currently adjusting. Similarly, both a music artist and a designer can at any time substitute one element of a composition by another, delete any elements, and add new ones. Most importantly, just as multitrack recording redefined the sound of popular music from the 1970s onward, once digital compositing became widely available during the 1990s it fundamentally changed the visual aesthetics of most moving images forms.

This discussion only scratched the surface of my subject in this section: layers and transparency. For instance, I have not analyzed the actual techniques of digital compositing and the fundamental concept of an alpha channel, which deserves a separate and detailed treatment. I have also not gone into the possible media histories leading to digital compositing nor examined its relationship to optical printing, video keying, and video effects technology of the 1980s. These histories and relationships were discussed in the "Compositing" chapter in *The Language of New Media* but from a different perspective than the one used here. At that time (1999) I was looking at compositing from the point of view of the questions of cinematic realism, practices of montage, and the construction of special effects in feature films. Today, however, it is clear to me that in addition to disrupting the regime of cinematic realism in favor of other visual aesthetics, compositing also had another, even more fundamental effect.

By the end of the 1990s digital compositing had become the basic operation used in creating *all* forms of moving images, and not only big budget features. So while it was originally developed as a technique for special effects in the 1970s and early 1980s,[25] compositing had a much broader effect on contemporary visual and media cultures beyond special effects. Compositing played the key role in turning the digital computer into a kind of experimental lab (or a Petri dish) where different media can meet and where their aesthetics and techniques can be combined to create new species. In short, digital compositing was essential in enabling the development of a new hybrid visual language of moving images—one that today we see everywhere.

Defined at first as a particular digital technique designed to integrate the two media of live action film and computer graphics in special effects sequences, compositing later became a "universal media integrator." And although compositing was originally created to support the aesthetics of cinematic realism, over time it actually had an opposite effect. Rather than forcing different media to fuse seamlessly, compositing led to the flourishing of numerous media hybrids where the juxtapositions between live and algorithmically generated, two dimensional and three dimensional, raster and vector media are made deliberately visible rather than being hidden.

After Effects interface: from "time-based" to "composition-based"

My thesis about media hybridity applies both to the cultural objects and the software used to create them. Just as the moving image media made by designers today mix the formats, assumptions, and techniques of different media, the toolboxes and interfaces of the software they use are also remixes. Let us again use After Effects as the case study to see how its interface remixes previously distinct working methods of different disciplines.

When moving image designers started to use compositing/animation software such as After Effects, its interface encouraged them to think about moving images in a fundamentally new way. Film and video editing systems and their computer simulations

[25] Thomas Porter and Tom Duff, "Compositing Digital Images," ACM *Computer Graphics* 18, no. 3 (July 1984): pp. 253–9.

that came to be known as non-linear editors (currently exemplified by Avid, Premiere, and Final Cut[26]) have conceptualized a media project as a sequence of shots organized in time. Consequently, while NLE (the standard abbreviation for non-linear video editing software) gave the editor many tools for adjusting the edits, they took for granted the constant of a film language that came from its industrial organization—that all frames have the same size and aspect ratio. This is an example of a larger trend. During the first stage of the development of cultural software, its pioneers were exploring the new possibilities of a computer metamedium going in any direction they were interested, since commercial use (with a notable exception of CAD) was not yet an option. However, beginning with the 1980s, a new generation of companies—Aldus, Autodesk, Macromedia, Adobe, and others—started to produce GUI-based software media authoring software aimed at particular industries: TV production, graphic design, animation, etc. As a result, many of the workflow principles, interface conventions and constraints of media technologies standard in these industries, were methodically re-created in software—even though the software medium itself has no such limitations. NLE software is a case in point. In contrast, from the beginning the After Effects interface put forward a new concept of moving image as a composition organized both in time and 2D space.

The center of After Effects' interface is a Composition conceptualized as a large canvas that acts as a background for visual elements which can have arbitrary sizes, proportions, and content (video, photos, abstract graphics, type, etc). When I first started using After Effects soon after it came out, I remember feeling shocked that the software did not automatically resize the graphics I dragged into the Composition window to make them fit the overall frame. The fundamental assumption of cinema that accompanied it throughout its whole history—that film consists of many frames which all have the same size and aspect ratio—was gone.

In film and video editing paradigms of the twentieth century, the minimal unit on which the editor works is a frame. S/he can change the length of an edit, adjusting where one film or video segment ends

[26] Compositing functionality was gradually added over time to most NLE systems, so today the distinction between Effects or Flame interfaces and Avid and Final Cut interfaces is less pronounced.

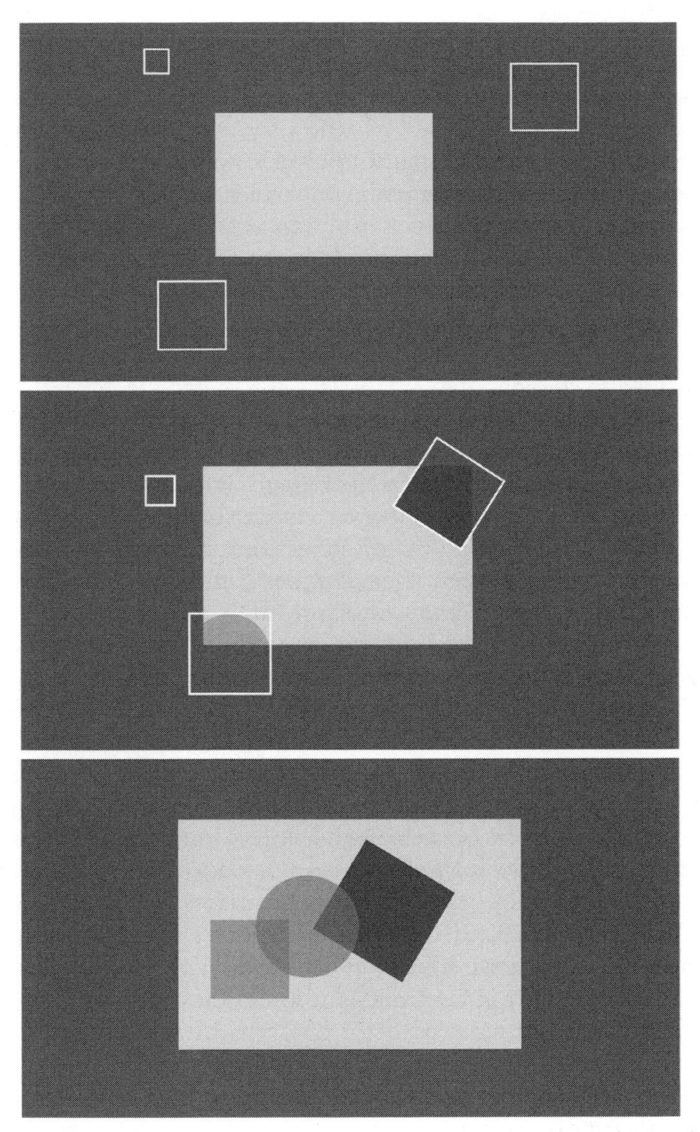

"The center of After Effects interface is a Composition conceptualized as a large canvas that acts as a background for visual elements which can have arbitrary sizes, proportions, and content. Each element can be individually accessed, manipulated, and animated." The illustration shows selected frames from a simple 2D animation, as they appear in the Composition panel.

and another begins, but s/he cannot directly modify the contents of a frame. The frame functions as a kind of "black box" that cannot be "opened." (This was the job for special effects departments and companies.) But in the After Effects interface, the basic unit is not a frame but a visual element placed in the Composition window. Each element can be individually accessed, manipulated, and animated. In other words, each element is conceptualized as an independent object. Consequently, a media composition is understood as a set of independent objects that can change over time. The very word "composition" is important in this context as it references 2D media (drawing, painting, photography, design) rather than filmmaking—i.e. space as opposed to time.

Where does the After Effects interface come from? Given that this software is commonly used to create animated graphics and visual effects, it is not surprising that its interface elements can be traced to three separate fields: animation, graphic design, and special effects. And because these elements are integrated in intricate ways to offer the user a new experience that cannot be simply reduced to a sum of the working methods already available in separate fields, it makes sense to think of the After Effects UI as an example of "deep remixability."

In twentieth-century cel animation production, an animator places a number of transparent cels on top of each other. Each cel contains a different drawing—for instance, a body of a character on one cel, the head on another cel, eyes on the third cel. Because the cels are transparent, the drawings get automatically "composited" into a single composition. While the After Effects interface does not use the metaphor of a stack of transparent cels directly, it is based on the same principle. Each element in the Composition window is assigned a "virtual depth" relative to all other elements. Together all elements form a virtual stack. At any time, the designer can change the relative position of an element within the stack, delete it, or add new elements.

We can also see a connection between the After Effects interface and another popular twentieth-century animation technique—stop motion. To create stop motion shots, puppets or any other 3D objects are positioned in front of a film camera and manually animated one step at a time. For instance, an animator may be adjusting the head of a character, progressively moving it from left to right in small discrete steps. After every step, the animator exposes one frame of film, then makes another adjustment, exposes another frame, and so on. (The twentieth-century animators and filmmakers who used this technique with great inventiveness include Wladyslaw Starewicz, Oskar Fischinger, Aleksandr Ptushko, Jiří Trnka, Jan Svankmajer, and the Brothers Quay.)

Just as in cel and stop-motion animation practices, After Effects does not make any assumptions about the size or positions of individual elements. Instead of dealing with standardized units of time—i.e. film frames containing fixed visual content—a designer now works with separate visual elements. An element can be a digital video frame, a line of type, an arbitrary geometric shape, etc. The finished work is the result of a particular arrangement of these elements in space and time. Consequently, a designer who uses After Effects can be compared to a choreographer who creates a dance by "animating" the bodies of dancers—specifying their entry and exit points, their trajectories through the space of the stage, and the movements of their bodies. (In this respect it is relevant that although the After Effects interface did not evoke this reference, another equally important 1990s software that was commonly used to design multimedia—Macromedia Director—did explicitly refer to the metaphor of the theatre stage in its UI.)

While we can link the After Effects interface to traditional animation methods as used by commercial animation studios, the working method put forward by software is closer to graphic design. In commercial animation studios of the twentieth century all elements—drawings, sets, characters, etc.—were prepared beforehand. The filming itself was a mechanical process. Of course, we can find exceptions to this industrial-like separation of labor in experimental animation practice where a film was usually produced by one person. This allowed a filmmaker to invent a film as he went along, rather than having to plan everything beforehand. A classic example of this is Oskar Fischinger's *Motion Painting 1* created in 1949. Fischinger made this 11-minute film by continuously modifying a painting and exposing film one frame at a time after each modification. This process took nine months. Because Fischinger was shooting on film, he had to wait a long time before seeing the results of his work. As the historian of abstract animation William Moritz writes, "Fischinger painted every day for over five months without being able to see how it was coming out on film, since he wanted to keep all the conditions, including film stock, absolutely consistent in order to avoid unexpected variations in quality of image."[27] In other words, in the case of this project by Fischinger, creating animation and seeing the result were even more separated than in a commercial animation process.

In contrast, a graphic designer works "in real time." As the designer introduces new elements, adjusts their locations, colors and other properties, tries different images, changes the size of the type, and so on, s/he can immediately see the result of his/her work.[28] After Effects adopts this working method by making the Composition window the center of its interface. Like a traditional designer, the After Effects user interactively arranges the elements in this window and can immediately see the result. In short, the After Effects interface makes filmmaking into a design process, and a film is re-conceptualized as a graphic design that can change over time.

As we saw when we looked at the history of cultural software, when physical or electronic media are simulated in a computer, we do not simply end up with the same media as before. By adding new properties and working methods, computer simulation fundamentally changes the identity of a given media. For example, in the case of "electronic paper" such as a Word document or a PDF file, we can do many things which were not possible with ordinary paper: zoom in and out of the document, search for a particular phrase, change fonts and line spacing, etc. Similarly, online interactive maps services provided by Google and Microsoft augment the traditional paper map in multiple and amazing ways.

A significant proportion of contemporary software for creating, editing, and interacting with media was developed in this way.

---

[27] Quoted in Michael Barrier, *Oskar Fischinger. Motion Painting No. 1*, http://www.michaelbarrier.com/Capsules/Fischinger/fischinger_capsule.htm

[28] Depending on the complexity of the project and the hardware configuration, the computer may or may not be able to keep pace with the designer's changes. Often a designer has to wait until the computer renders everything in frame after s/he makes changes. However, since s/he has control over this rendering process, s/he can instruct After Effects to show only outlines of the objects, to skip some layers, etc.—thus giving the computer less information to process and allowing for real-time feedback. While a graphic designer does not have to wait until a film is developed or a computer has finished rendering the animation, the design has its own "rendering" stage—making proofs. With both digital and offset printing, after the design is finished, it is sent to the printer who produces the test prints. If the designer finds any problems such as incorrect colors, s/he adjusts the design and then asks for proofs again.

Already existing media technologies were simulated in a computer and augmented with new properties. But if we consider media authoring software such as Maya (3D modeling and computer animation) or After Effects (motion graphics, compositing and visual effects), we encounter a different logic. These software applications do not simulate any single physical media that existed previously. Rather, they borrow from a number of different media, combining and mixing their working methods and specific techniques. (And, of course, they also add new capabilities specific to computers—for instance, the ability to automatically calculate the intermediate values between a number of key frames.) For example, 3D modeling software mixes form-making techniques which were previously "hardwired" to different physical media: the ability to change the curvature of a rounded form as though it were made from clay, the ability to build a complex 3D object from simple geometric primitives the way buildings were constructed from identical rectangular bricks, cylindrical columns, pillars, etc.

Similarly, as we saw, the After Effects original interface, toolkit, and workflow draws upon the techniques of animation and graphic design. (We can also find traces of filmmaking and 3D computer graphics.) But the result is not simply a mechanical sum of all elements that came from earlier media. Rather, as software remixes the techniques and working methods of the various media they simulate, the results are new interfaces, tools, and workflows with their own distinct logic. In the case of After Effects, the working method that it puts forward is neither animation nor graphic design nor cinematography, even though it draws from all these fields. It is a new way to make moving image media. Similarly, the visual language of media produced with this and similar software is also different from the languages of moving images that existed previously.

Consequently, the Velvet Revolution unleashed by After Effects and other software did not simply make more commonplace the animated graphics that artists and designers such as John and James Whitney, Norman McLaren, Saul Bass, Robert Abel, Harry Marks, Richard and Robert Greenberg were previously creating using stop motion animation, optical printing, video effects hardware of the 1980s, and other custom techniques and technologies. Instead it led to the emergence of numerous new visual aesthetics that did not exist before. And if the common feature of these aesthetics is "deep remixability," it is not hard to see that it mirrors the "deep remixability" in the After Effects UI.

3D space as a media design platform

As I was researching what users and industry reviewers had been saying about After Effects, I came across a somewhat condescending characterization of this software as "Photoshop with keyframes." I think that this characterization is actually quite useful.[29] Think about all the different ways of manipulating images available in Photoshop and the degree of control provided by its multiple tools. Think also about Photoshop's concept of a visual composition as a stack of potentially hundreds of layers, each with its own transparency setting and multiple alpha channels. If we are able to animate such a composition and continue using Photoshop's tools to adjust visual elements over time on all layers independently, this indeed constitutes a new paradigm for creating moving images. And this is what After Effects and other animation, visual effects, and compositing software make possible today.[30] And while the idea of working with a number of layers placed on top of each other itself is not new—consider traditional cel animation, optical printing, video switchers, photo-collage, graphic design—going from a few non-transparent layers to hundreds and even thousands of layers, each with its controls, fundamentally changes not only how a moving image looks but also what it can say. From being a special effect reserved for particular shots, 2D compositing became a part of the standard animation and video editing interface.

But innovative as the 2D compositing paradigm was, by the beginning of the 2000s it was supplemented by a new one: 3D compositing. If 2D compositing can be thought as an extension

---

[29] Soon after the initial release of After Effects in January 1993, CoSA (the company that produced this software) was purchased by Aldus, which in turn was purchased by Adobe—which was already selling Photoshop.

[30] Photoshop and After Effects were designed originally by different teams at different times, and even after both products were purchased by Adobe (it released Photoshop in 1989 and After Effects in 1995), it took Adobe a number of years to build close links between After Effects and Photoshop eventually making it easy to move assets between the two programs.

of already familiar media techniques, the new paradigm does not come from any previous physical or electronic media. Instead, it takes the new born-digital media which was invented in the 1960s and matured by the early 1990s—interactive 3D computer graphics and animation—and transforms it into a general platform for moving media design.

The language used in the professional production milieu today reflects an implicit understanding that 3D graphics is a new medium unique to a computer. When people use the terms "computer visuals," "computer imagery," or "CGI" (which is an abbreviation for "computer generated imagery") everybody understands that they refer to 3D graphics as opposed to other image sources such as "digital photography." I think of 3D computer graphics as a new media—as opposed to considering it an extension of architectural drafting, projection geometry, or set making—because they offer a new method for representing three-dimensional reality: both objects that already exist and objects that are only imagined. This method is fundamentally different from what has been offered by the main representational media of the industrial era: lens-based capture (still photography, film recording, video) and audio recording. With 3D computer graphics, we can represent a three-dimensional structure of the world, versus capturing only a perspectival image of the world, as in lens-based recording. We can also manipulate our representation using various tools with ease and precision—qualitatively different from the much more limited "manipulability" of a model made from any physical material (although nanotechnology promises to change this in the future). And, as contemporary architectural aesthetics makes clear, 3D computer graphics is not simply a faster way of working with geometric representations like the plans and cross-sections used by draftsmen for centuries. When generations of young architects and architectural students started to systematically work with 3D modeling and animation software such as Alias in the middle of the 1990s, the ability to directly manipulate a 3D shape (rather than only dealing with its projections as in traditional drafting) quickly led to a whole new language of complex non-rectangular curved forms. In other words, architects working with the media of 3D computer graphics started to imagine different things than their predecessors who used pencils, rules, and drafting tables.

When the Velvet Revolution of the 1990s made it possible to easily combine multiple media sources in a single moving image sequence using multi-layer interface of After Effects, CGI was added to the mix. Today, 3D models are routinely used in media compositions created in After Effects and similar software, along with all other media sources. But in order to be a part of the mix, these models need to be placed on their own 2D layers and thus treated as 2D images. This was the original After Effects paradigm: all image media can meet as long as they are reduced to 2D.[31]

In contrast, in the 3D compositing paradigm all media types are placed within a single 3D space. One advantage of this representation is that since 3D space is "native" to 3D computer graphics, 3D models can stay as they are, i.e. three-dimensional. An additional advantage is that the designer can now use all the techniques of virtual cinematography as developed in 3D computer animation. S/he can define different kinds of lights, fly the virtual camera around and through the image planes using any trajectory, and use depth of field and motion blur effects.[32]

While 3D computer-generated models already "live" in this space, how do you bring in two-dimensional visual elements—video, digitized film, typography, drawn images? If 2D compositing paradigm treated everything as 2D images—including 3D computer models—3D compositing treats everything as 3D. So while two-dimension elements do not inherently have a third dimension, it has to be added to enable these elements to enter the three-dimensional space. To do that, a designer places flat cards in this

---

[31] I say "original" because the later version of After Effects added the ability to work with 3D layers.

[32] If 2D compositing can be understood as an extension of twentieth-century cel animation, where a composition consists of a stack of flat drawings, the conceptual source of the 3D compositing paradigm is different. It comes out of the work on integrating live action footage and CGI done in the 1980s in the context of feature films production. Both film director and computer animator work in a three-dimensional space: the physical space of the set in the first case, the virtual space as defined by 3D modeling software in the second case. Therefore conceptually it makes sense to use three-dimensional space as a common platform for the integration of these two worlds. It is not accidental that NUKE, one of today's leading programs for 3D compositing was developed in-house at Digital Domain, which was co-founded in 1993 by James Cameron, the Hollywood director who has systematically advanced the integration of CGI and live action in films such as *The Abyss* (1989), *Terminator 2* (1991), and *Titanic* (1997).

"In the 3D compositing paradigm all media types are placed within a single 3D space." This illustration uses the animation from earlier, but adds the third dimension. We positioned each element at a different depth, and added a light, casting shadows.

space in particular locations, and situates two-dimensional images on these cards. Now, everything lives in a common 3D space. This condition enables "deep remixability" between techniques which I have illustrated using the example of "Go" video. The techniques of drawing, photography, cinematography, and typography which go into capturing or creating two-dimensional visual elements can now "play" together with all the techniques of 3D computer animation (virtual camera moves, controllable depth of field, variable lens, etc.)

In 1995 I wrote the article 'What is Digital Cinema?' The article asked how the changes in moving image production I was witnessing (I was living in Los Angeles at that time, which made it easy to follow what was happening in Hollywood) were affecting the meaning of "cinema." In that article I proposed that the logic of hand-drawn animation, which throughout the twentieth century was marginal in relation to cinema, became dominant in a software era. Because software allows the designer to manually manipulate any image regardless of its source as though it was drawn in the first place, the ontological differences between different image media become irrelevant. Both conceptually and practically, they are all reduced to hand-drawn animation.

By default, After Effects and other animation/video editing/2D compositing software treats a moving image project as a stack of layers. Therefore, I can extend my original argument and propose that animation logic also moves from the marginal to the dominant position in yet another way. The paradigm of a composition as a stack of separate visual elements as practiced in cel animation becomes the default way of working with all images in a software environment—regardless of their origin and final output media. In other words, a "moving image" is now understood as a composite of layers of imagery—rather than as a still flat picture that only changes in time, as it was the case for most of the twentieth century. In the word of animation, editing, and compositing software, such a "single layer image" becomes an exception.

The emergence of the *3D compositing paradigm* can be also seen as following this logic of historical reversal. The new representational structure as developed within the computer graphics field—a 3D virtual space containing 3D models—has gradually moved from a marginal to a dominant role. In the 1970s and 1980s computer graphics were used only occasionally in a dozen feature films such as *Alien* (1979), *Tron* (1981), *The Last Starfighter* (1984), and *The Abyss* (1989), and selected television commercials and broadcast graphics. But by the beginning of the 2000s, the representation structure of computer graphics, i.e. a 3D virtual space, came to function as an umbrella for all other image types regardless of their origin. An example of an application which implements this paradigm is Flame, enthusiastically described by one user as "a full 3D compositing environment into which you can bring 3D models, create true 3D text and 3D particles, and distort layers in 3D space."[33]

This does not mean that 3D animation itself became visually dominant in moving image culture, or that the 3D structure of the space within which media compositions are now routinely constructed is necessarily made visible (usually it is not). Rather, the way 3D computer animation organizes visual data—as objects positioned in a Cartesian space—became the way to work with all moving image media. As already stated above, a designer positions all the elements which go into a composition—2D animated sequences, 3D objects, particle systems, video and digitized film sequences, still images and photographs—inside the shared 3D virtual space. There these elements can be further animated, transformed, blurred, filtered, etc. So while all moving image media has been reduced to the status of hand-drawn animation in terms of their manipulability, we can also state that all media have become layers in 3D space. In short, the new media of 3D computer animation has "eaten up" the dominant media of the industrial age—lens-based photo, film and video recording.

Having discussed how software has redefined the concept of a "moving image" as a composite of multiple layers, this is a good moment to pause and consider other possible ways software changed this concept. When cinema in its modern form was born in the end of the nineteenth century, the new medium was understood as an extension of already familiar one—that is, as a photographic image which was now moving. This understanding can be found in the press accounts of the day and also in one of the official names given to the new medium—"moving pictures." On the material

[33] Alan Okey, post to forums.creativecow.net, December 28, 2005, http://forums.creativecow.net/cgi-bin/dev_read_post.cgi?forumid=154&postid=855029

level, a film indeed consisted of separate photographic frames. When they quickly replace each other, this creates the effect of motion for the viewer. So the concept used to understand cinema indeed fit in with the structure of the medium.

But is this concept still appropriate today? When we record video and play it, we are still dealing with the same structure: a sequence of frames. But for professional media designers, the terms have changed. The importance of these changes is not just academic and purely theoretical—because designers understand their media differently, they are creating films and sequences that also look very different from twentieth-century cinema or animation.

Consider those new ways of creating moving images (which I referred to as a new paradigms) that we have discussed thus far. (Although theoretically they are not necessarily all compatible with each other, in production practice these different paradigms are used in a complementary fashion.) A "moving image" became a hybrid which can combine all different visual media invented so far—rather than holding only one kind of data such as camera recording, hand drawing, etc. Rather than being understood as a singular flat plane—the result of light focused by the lens and captured by the recording surface—it is now understood as a stack of a potentially infinite number of separate layers. And rather than "time-based," it becomes "composition-based," or "object—oriented." That is, instead of being treated as a sequence of frames arranged in time, a "moving image" is now understood as a two-dimensional composition that consists of a number of objects that can be manipulated independently. Alternatively, if a designer uses 3D compositing, the conceptual shift is even more dramatic: instead of editing "images," s/he is working in a virtual three-dimensional space that holds both CGI and lens-recorded flat image sources.

Of course, frame-based representation did not disappear—it became simply a recoding and output format rather than the space where a film is being put together. And while the term "moving image" can be still used as an appropriate description for how the output of a production process is experienced by the viewers, it is no longer captures how the designers think about what they create. Because their production environment—workflow, interfaces, and tools—has changed so much, they are thinking today very differently than twenty years ago.

If we focus on what the different paradigms summarized above have in common, we can say that filmmakers, editors, special effects artists, animators, and motion graphics designers are all working on *a composition in 2D or a 3D space that consists of a number of separate objects*. The spatial dimension has become as important as the temporal dimension. From the concept of a "moving image" understood as a sequence of static photographs we have moved to a new concept: *a modular media composition*. And while a person who directs a feature or a short film that is centered around actors and live action can be still called a "filmmaker," in all other cases where most of the production takes place in a software environment, it is more appropriate to call the person a "designer." This is yet another fundamental change in the concept of "moving images": today more often than not they are not "captured," "directed," or "animated." Instead, they are "designed."

Import/export: design workflow

In our discussions of the After Effects interface and workflow, as well as the newer paradigm of 3D compositing, we have already come across the crucial aspect of software-based media production process. Until the arrival of software-based tools in the 1990s, combining different types of time-based media together was either time consuming, expensive, or in some cases simply impossible. Software tools such as After Effects have changed this situation in a fundamental way. Now a designer can import different media into their composition with just a few mouse clicks.

However, the contemporary software-based design of moving images—or any other design process, for that matter—does not simply involve combining elements from different sources within a single application. In this section we will look at the whole workflow typical of contemporary design—be it the design of moving images, still illustrations, 3D objects and scenes, architecture, music, websites, or any other media. (Most of the details of software-based production of moving images which I have already presented also applies to the graphic design of still images and layouts for print, the web, packaging, physical spaces, mobile devices, etc. However, in this section I want to make this explicit.

Therefore the examples below will include not only moving images, but also graphic design.)

Although "import/export" commands appear in most modern media authoring and editing software applications running under a GUI, at first sight they do not seem to be very important for understanding software culture. When you "import," you are neither authoring new media nor modifying media objects, nor accessing information across the globe, as in web browsing. All these commands allow you to do is to move data around between different applications. In other words, they make data created in one application compatible with other applications. And that does not look so important.

Think again. What is the largest part of the economy of the greater Los Angeles area? It is not entertainment. From movie production to museums and everything in between, entertainment only accounts for 15 percent. It turns out that the largest part of the economy is the import/export business—more than 60 percent. More generally, one commonly evoked characteristic of globalization is greater connectivity—places, systems, countries, organizations etc. becoming connected in more and more ways. And connectivity can only happen if you have certain level of compatibility: between business codes and procedures, between shipping technologies, between network protocols, between computer file formats, and so on.

Let us take a closer look at import/export commands. As I will try to show below, these commands play a crucial role in software culture, and in particular in media design—regardless of what kind of project a design is working on.

Before they adopted software tools in the 1990s, filmmakers, graphic designers, and animators used completely different technologies. Therefore, as much as they were influenced by each other or shared the same aesthetic sensibilities, they inevitably created different-looking images. Filmmakers used camera and film technology designed to capture three-dimensional physical reality. Graphic designers were working with offset printing and lithography. Animators were working with their own technologies: transparent cels and an animation stand with a stationary film camera capable of making exposures one frame at a time as the animator changed cels and/or moved background.

As a result, twentieth-century cinema, graphic design, and animation (I am talking here about standard animation techniques used by most commercial studios) developed distinct artistic languages and vocabularies both in terms of form and content. For example, graphic designers worked with a two dimensional space, film directors arranged compositions in three-dimensional space, and cel animators worked with "two-and-a-half" dimensions. This holds for the overwhelming majority of works produced in each field, although of course exceptions do exist. For instance, Oskar Fischinger made one abstract film that consisted of simple geometric objects moving in an empty space—but as far as I know this is the only film in the history of abstract pre-digital animation that takes place in three-dimensional space.

Differences in technology influenced what kind of content would appear in different media. Cinema showed "photorealistic" images of nature, built environments and human forms articulated by special lighting. Graphic designs featured typography, abstract graphic elements, monochrome backgrounds, and cutout photographs. And cartoons presented hand-drawn flat characters and objects animated over hand-drawn (but more detailed) backgrounds. The exceptions are rare. For instance, while architectural spaces frequently appeared in films because directors could explore their three—dimensionality in staging scenes, they practically never appeared in animated films in any detail—until animation studios started using 3D computer animation.

Why was it so difficult to cross boundaries? For instance, in theory one could imagine making an animated film in the following way: printing a series of slightly different graphics designs and then filming them as though they were a sequence of animated cels. Or a film where a designer simply made a series of hand drawings that used the exact vocabulary of graphic design and then filmed them one by one. And yet, to the best of my knowledge, such a film was never made. What we find instead are many abstract animated films that have a certain connection to various styles of abstract painting. For example, Oskar Fischinger's films and paintings share certain forms. We can also find abstract films and animated commercials and movie titles that have a connection to graphic design aesthetics popular around the same time. For instance, some moving image sequences made by motion graphics pioneer Pablo Ferro in the 1960s display psychedelic aesthetics which can also be found in posters, record covers, and other works of graphic design in the same period.

And yet despite these connections, works in different media never used exactly the same visual language. One reason is that projected film could not adequately show the subtle differences between typeface sizes, line widths, and gray-scale tones crucial for modern graphic design. Therefore, when the artists were working on abstract art films or commercials that adopted design aesthetics (and most major twentieth-century abstract animators worked both on their personal films and commercials), they could not simply expand the language of a printed page into the time dimension. They essentially had to invent a parallel visual language that used bold contrasts, more easily readable forms, and thick lines—that, because of their thickness, were in fact no longer lines but shapes.

Although the limitations in resolution and contrast of film and television image in comparison to a printed page contributed to the distance between the languages used by abstract filmmakers and graphic designers for the most of the twentieth century, ultimately I do not think it was the decisive factor. Today the resolution, contrast and color reproduction between print, computer screens, television screens, and the screens of mobile phones are also substantially different—and yet we often see exactly the same visual strategies deployed across these different display media. If you want to be convinced, leaf through any book or a magazine on contemporary 2D design (i.e., graphic design for print, broadcast, and the web). When you look at pages featuring the works of a particular designer or a design studio, in most cases it is impossible to identify the origins of the images unless you read the captions. Only then do you find which image is a poster, which one is a still from a music video, and which one is a magazine editorial.

I am going to use *Graphic Design for the 21st Century: 100 of the World's Best Graphic Designers* (Taschen, 2001) for my examples because by 2001, the changes I describe had already taken place. Peter Anderson's design showing a line of type against a cloud of hundreds of little letters in various orientations turns out to be the frames from the title sequence for a Channel 4 documentary. His other design which similarly plays on the contrast between jumping letters in a larger font against irregularly cut planes made from densely packed letters in much smaller fonts, turns to be a spread from IT Magazine. Since the first design was made for broadcast while the second was made for print, we would expect that the first design would employ bolder forms—however, both designs use the same scale between big and small fonts, and feature texture fields composed from hundreds of words in such a small font that they are clearly not intended to be read. A few pages later we encounter a design by Philippe Apeloig that uses exactly the same technique and aesthetics as Anderson used. In this case, tiny lines of text positioned at different angles form a 3D shape floating in space. On the next page another design by Apeloig creates a field in perspective—made not from letters but from hundreds of identical abstract shapes.

These designs rely on the software's ability (or on the designer being influenced by software use and recreating what s/he did with software manually) to treat text as any graphical primitive and to easily create compositions made from hundreds of similar or identical elements positioned according to some pattern. And since an algorithm can easily modify each element in the pattern, changing its position, size, color, etc., instead of the completely regular grids of modernism we see more complex structures that are made from many variations of the same element. This strategy is explored particularly imaginatively in Zaha Hadid's designs such as the Louis Vuitton "Icone Bag" 2006, and in her urban master plans for Singapore and Turkey, which use what Hadid called a "variable grid."

Each designer included in the book was asked to provide a brief statement to accompany the portfolio of his/her work, and Lust Studio has chosen the phrase "Form-follows-process" as their motto. So what is the nature of the design process in the software age and how does it influence the forms we see today around us?

If you are practically involved in design or art today, you already know that contemporary designers use the same small set of software tools to design just about everything. I have already named them repeatedly, so you know the list: Photoshop, Illustrator, Flash, Maya, etc. However, the crucial factor is not the tools themselves but the workflow process, enabled by "import" and "export" operations and related methods ("place," "insert object," "subscribe," "smart object," etc.), that ensure coordination between these tools.

When a particular media project is being put together, the software used at the final stage depends on the type of output media and the nature of the project—After Effects for motion graphics

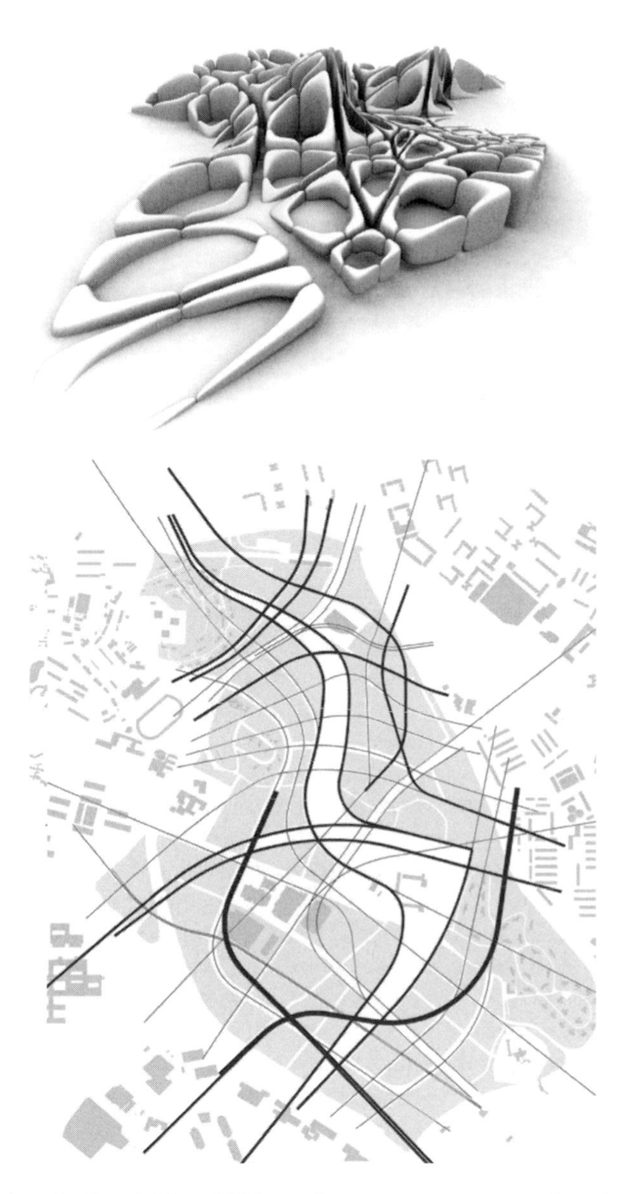

Examples of a "variable grid" in architecture: two master plans by Zaha Hadid. Top: Kartal Pendik, Istanbul, 2006. Bottom: One North, Singapore, 2001.

projects and video compositing, Illustrator for print illustrations, InDesign for multi-page designs, Flash for interactive interfaces and web animations, 3ds Max or Maya for 3D computer models and animations, and so on. But these programs are rarely used alone to create a media design from start to finish. Typically, a designer may create elements in one program, import them into another program, add elements created in yet another program, and so on. This happens regardless whether the final product is an illustration for print, a website, or a motion graphics sequence, whether it is a still or a moving image, interactive or non-interactive, etc.

The very names which software companies give to the products for media design and production refer to this defining characteristic of software-based design process. Since 2005, Adobe has been selling its different media authoring applications bundled together under the name "Adobe Creative Suite." Among the subheadings and phrases that were used to accompany this brand name, one in particular is highly meaningful in the context of our discussion: "Design Across Media." This phrase accurately describes both the capabilities of the applications collected in a suite and their actual use in the real world. Each of the key applications collected in the suite—Photoshop, Illustrator, InDesign, Flash, Dreamweaver, After Effects, Premiere—has many special features geared for producing a design for particular output media. Illustrator is set up to work with professional-quality printers; After Effects and Premiere can output video files in a variety of standard video formats such as HDTV; Dreamweaver supports programming and scripting languages to enable creation of sophisticated and large-scale dynamic websites. But while a design project is finished in one of these applications, most other applications in Adobe Creative Suite will be used in the process to create and edit its various elements. This is one of the ways in which Adobe Creative Suite enables "design across media." The compatibility between applications also means that the elements (called in professional language "assets") can be later re-used in new projects. For instance, a photograph edited in Photoshop can first be used in a magazine ad and later put in a video, a website, etc. Or, the 3D models and characters created for a feature film can be reused for a video game based on the film. This ability to re-use the same design elements for very different projects is very important because of the widespread practice in creative industries to create products across the range of media which share the same images, designs, characters, narratives, etc. An advertising campaign often works "across media" including web ads, TV ads, magazine ads, billboards, etc. And if turning movies into games and games into movies has been already popular in Hollywood since the early 1990s, a new trend since approximately the middle of the 2000s is to create a movie, a game, a website or maybe other media products at the same time—and have all the products use the same digital assets both for economic reasons and to assure aesthetic continuity between these products. Thus, a studio may create 3D backgrounds and characters and put them in both a movie and a game, both of which will be released simultaneously. If media authoring applications were not compatible, such a practice would simply not be possible.

All these examples illustrate the intentional reuse of design elements "across media." However, the compatibility between media authoring applications also has a much broader and unintentional effect on contemporary aesthetics. Given the production workflow I have just described, we may expect that the same visual techniques and strategies will also appear in all types of media projects designed with software, without this being consciously planned for. We may also expect that this will happen on a much more basic level. This is indeed the case. *The same software-enabled design strategies, the same software-based techniques, and the same software-generated iconography are now found across all types of media, all scales, and all kinds of projects.*

We have already encountered a few concrete examples. For instance, the three designs by Peter Anderson and Philippe Apeloig done for different media use the same basic computer graphic technique: automatic generation of a repeating pattern while varying the parameters that control the appearance of each element making up the pattern—its size, position, orientation, curvature, etc. (The general principle behind this technique can also be used to generate 3D models, animations, textures, make plants and landscapes, etc. It is often referred to as "parametric design," or "parametric modeling.") The same technique is also used by Hadid's studio for the Louis Vuitton "Icone Bag". In another example, which will be discussed below, Greg Lynn used particle systems technique—which at that time was normally used to simulate fire, snow, waterfalls, and other natural phenomena in cinema—to generate the forms of a building.

To use the biological metaphor, we can say that compatibility between design applications creates very favorable conditions for the propagation of media DNA between species, families, and classes. And this propagation happens on all levels: the whole design, parts of a design, the elements making up the parts, and the "atoms" that make up the elements. Consider the following hypothetical example of propagation on a lower level. A designer can use Illustrator to create a 2D smooth curve (called in the computer graphics field a "spline.") This curve becomes a building block that can be used in any project. It can form part of an illustration or a book design. It can be imported into an animation program where it can be set to motion, or imported into 3D program where it can be extruded in 3D space to define a solid object.

Over time software manufacturers worked to develop tighter ways of connecting their applications, in order to make it easier to move elements from one application to another. Over the years, it became possible to move a whole project between applications without losing anything (or almost anything). For example, in describing the integration between Illustrator CS3 and Photoshop CS3, Adobe's website states that a designer can "Preserve layers, layer comps, transparency, and editable files when moving files between Photoshop and Illustrator."[34] Another important development has been the concept that Microsoft Office calls "linked objects." If you link all or part of one file to another file (for instance, linking an Excel document to a PowerPoint presentation), anytime information changes in the first file it automatically gets updated in the second file. Many media applications implement this feature. To use the same example of Illustrator CS3, a designer can "Import Illustrator files into Adobe Premiere Pro software, and then use Edit Original command to open the artwork in Illustrator, edit it, and see your changes automatically incorporated into your video project."[35]

Each type of program used by media designers—3D graphics, vector drawing, image editing, animation, compositing—excel at particular design operations, i.e. particular ways of creating design elements or modifying already existing elements. These

---

[34] http://www.adobe.com/products/illustrator/features/allfeatures/ (August 30, 2008).
[35] *Ibid.*

operations can be compared to the different types of blocks of a LEGO set. You can create an infinite number of projects by using the limited number of block types provided in the set. Depending on the project, these block types will have different functions and appear in different combinations. For example, a rectangular block may become part of the tabletop, part of the head of a robot, etc.

A design workflow that uses a small number of compatible software programs works in a similar way—with one important difference. The building blocks used in contemporary design are not only different kinds of visual elements one can create—vector patterns, 3D objects, particle systems, etc.—but also *various ways of modifying these elements*: blur, skew, vectorize, change transparency level, extrude, etc. This difference is crucial. If media creation and editing software did not include these and many other modification operations, we would see an altogether different visual language at work today. We would see "multimedia," i.e. designs that simply combine elements from different media. Instead, we see "deep remixability"—the "deep" interactions between working methods and techniques of different media within a single project.

In a "crossover" use, the techniques which were previously specific to one media are applied to other media types (for example, a lens blur filter). This often can be done within a single application—for instance, applying After Effects' blur filter to a composition which can contain graphic elements, video, 3D objects, etc. And being able to move a whole project or its elements between applications opens more possibilities because each application offers many unique techniques not available in other applications. As the media data travels from one application to the next, it is being transformed and enhanced using the operations offered by each application. For example, a designer can take the project s/he has been editing in Adobe Premiere and import it in After Effects where s/he can use advanced compositing features of that program. S/he can then import the result back into Premiere and continue editing. Or s/he can create artwork in Photoshop or Illustrator and import it into Flash where it can be animated. This animation can be then imported into a video editing program and combined with video. And so on.

The production workflow specific to the software era that I have just illustrated has two major consequences. Its first result is the visual aesthetics of hybridity that dominates contemporary design universe. The second is the use of the same techniques and strategies across this universe—regardless of the output media and type of project.

As I have already stated more than once, a typical design today combines techniques coming from multiple media. We are now in a better position to understand why this is the case. As a designer works on a project, s/he combines the results of the operations specific to different software programs that were originally created to imitate work with different physical media (Illustrator was created to make illustrations, Photoshop to edit digitized photographs; Premiere to edit video, etc.) While these operations continue to be used in relation to their original media, most of them are now also used as part of the workflow on any design job.

The essential condition that enables this new design logic and the resulting aesthetics is compatibility between files generated by different programs. In other words, "import," "export" and related functions and commands of graphics, animation, video editing, compositing and modeling software are historically more important than the individual operations these programs offer. The ability to combine raster and vector layers within the same image, place 3D elements into a 2D composition and vice versa, etc. is what enables the production workflow and its reuse of the same techniques, effects, and iconography across different media.

The consequences of this compatibility between software and file formats, which was gradually achieved during the 1990s, are hard to overestimate. Besides the hybridity of modern visual aesthetics and the reappearance of exactly the same design techniques across all output media, there are also other effects. For instance, the whole field of motion graphics as it exists today came into existence to a large extent because of the integration between vector drawing software, specifically Illustrator, and animation/compositing software such as After Effects. A designer typically defines various composition elements in Illustrator and then imports them into After Effects where they are animated. This compatibility did not exist when the initial versions of different media authoring and editing software became available in the 1980s. It was gradually added in particular software releases. But when it was achieved around the middle of the 1990s,[36] within a few years the whole language of contemporary graphic design was fully imported into the moving image area—both literally and metaphorically.

In summary, the compatibility between graphic design, illustration, animation, video editing, 3D modeling and animation, and visual effects software plays the key role in shaping the visual and spatial forms of the software age. On the one hand, never before have we witnessed such a variety of forms as today. On the other hand, exactly the same techniques, compositions and iconography can now appear in any media.

Variable form

As the films of Blake and Murata discussed earlier illustrate, in contemporary motion graphics the transformations often affect the frame as a whole. In contrast to twentieth-century animation, everything inside the frame keeps changing: visual elements, their transparency, the texture of the image, etc. In fact, if something stays the same for a while, that is an exception rather than the norm.

Such *constant change on many visual dimensions* is another key feature of motion graphics and design cinema produced today. Just as we did in the case of media hybridity, we can connect this preference for constant change to the particulars of software used in media design.

Digital computers allow us to represent any phenomenon or structure as a set of variables. In the case of design and animation software, this means that all possible forms—visual, temporal, spatial, interactive—are similarly represented as sets of variables that can change continuously. This new logic of form is deeply encoded in the interfaces of software packages and the tools they provide.

Consider again the After Effects interface. To create an animation, a designer adds a number of elements to the composition and then animates them. Each new element shows up in the interface as a

[36] In 1995, After Effects 3.0 enabled importing Illustrator files and Photoshop files as compositions. http://en.wikipedia.org/wiki/Adobe_After_Effects

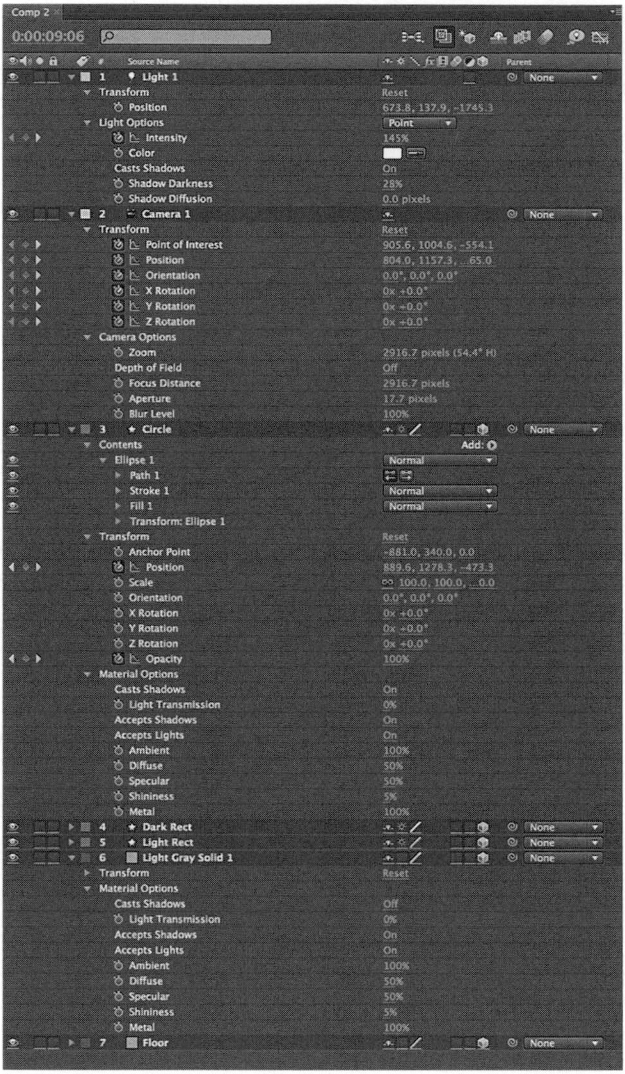

Comp 2
0:00:09:06

Source Name	Value	Parent
1 Light 1		None
Transform	Reset	
Position	673.8, 137.9, -1745.3	
Light Options	Point	
Intensity	145%	
Color		
Casts Shadows	On	
Shadow Darkness	28%	
Shadow Diffusion	0.0 pixels	
2 Camera 1		None
Transform	Reset	
Point of Interest	905.6, 1004.6, -554.1	
Position	804.0, 1157.3, -65.0	
Orientation	0.0°, 0.0°, 0.0°	
X Rotation	0x +0.0°	
Y Rotation	0x +0.0°	
Z Rotation	0x +0.0°	
Camera Options		
Zoom	2916.7 pixels (54.4° H)	
Depth of Field	Off	
Focus Distance	2916.7 pixels	
Aperture	17.7 pixels	
Blur Level	100%	
3 Circle		None
Contents	Add:	
Ellipse 1	Normal	
Path 1		
Stroke 1	Normal	
Fill 1	Normal	
Transform: Ellipse 1		
Transform	Reset	
Anchor Point	-881.0, 340.0, 0.0	
Position	889.6, 1278.3, -473.3	
Scale	100.0, 100.0, 0.0	
Orientation	0.0°, 0.0°, 0.0°	
X Rotation	0x +0.0°	
Y Rotation	0x +0.0°	
Z Rotation	0x +0.0°	
Opacity	100%	
Material Options		
Casts Shadows	On	
Light Transmission	0%	
Accepts Shadows	On	
Accepts Lights	On	
Ambient	100%	
Diffuse	50%	
Specular	50%	
Shininess	5%	
Metal	100%	
4 Dark Rect		None
5 Light Rect		None
6 Light Gray Solid 1		None
Transform	Reset	
Material Options		
Casts Shadows	Off	
Light Transmission	0%	
Accepts Shadows	On	
Accepts Lights	On	
Ambient	100%	
Diffuse	50%	
Specular	50%	
Shininess	5%	
Metal	100%	
7 Floor		None

"To create an animation, a designer adds a number of elements to the composition and then animates them. Each new element shows up in the interface as a list of parameters. Animating any parameter is equally easy, and it only takes a few clicks." The illustration shows a part of After Effects Timeline panel for the 3D animation illustrated earlier. The Timeline contains one light, one camera, three shapes (two rectangles and a circle), and a horizontal plane.

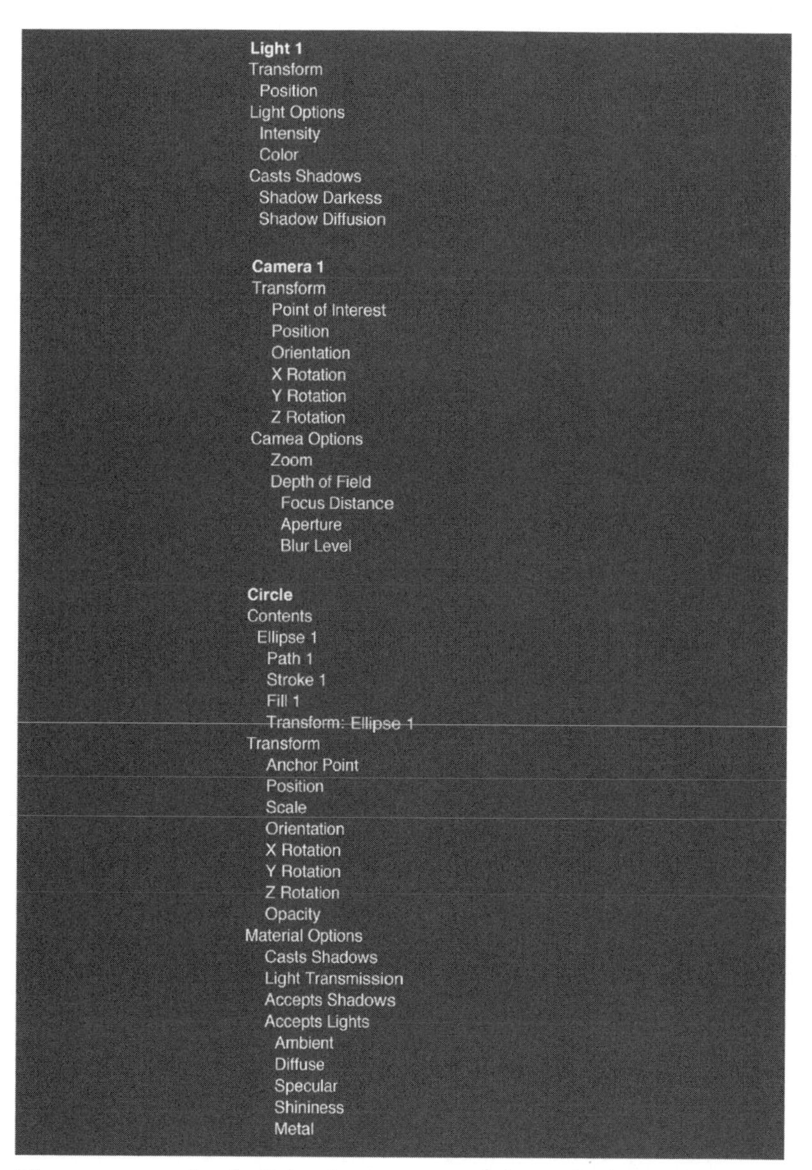

The parameters for the light, the camera, and the circle shape as they appear in After Effects interface. Each parameter can be animated separately. (To create this illustration, we redrew a part of the After Effects Timeline panel screenshot that appears on the opposite page to only show the parameter names.)

list of parameters, each with its animation channel. Depending on the element type, the parameters range from a few to a dozen or more. For example, for lights, their parameters are intensity, color, shadow darkness and shadow diffusion. If an element is a camera, the parameters include its point of interest, position, orientation, and rotation. For shapes, the list of parameters is particularly long: position, scale, orientation, opacity, material properties including cast shadows, light transmission, ambient, diffuse, and specular qualities, shininess, and others.

Animating any parameter only takes a few clicks. The process works the same regardless or the element and parameter type. Thus, it is equally easy to change over time the position of a shape, its color, or the intensity of a light.

In contrast to twentieth-century animation, After Effects does not privilege movement of two-dimensional objects and characters. Accordingly, After Effects Help defines "animation" as "change over time"—without specifying what can change.[37] The answer is provided by the interface itself. Its design suggests that you can animate hundreds of visual characteristics (and if we also consider that most filters can also be animated, the list goes into thousands).

Because the software interface makes directly visible every parameter for every object in the composition, assigning each its own channel on the timeline, it literally invites the designer to start animating them. You are invited to start moving and rotating objects, changing their opacity, colors, and so on. The same logic extends to the camera and the lights. If you add a light to the composition, this immediately creates half a dozen new animation channels describing light's color, position, orientation, intensity, and shadow properties, each with its own timeline channel. (Other 2D and 3D animation and layer-based compositing software packages all share the same interface principles.[38])

---

[37] http://help.adobe.com/en_US/AfterEffects/9.0/ (November 7, 2012).

[38] Node-based compositing software uses a different interface principle: each element in a composition is represented as node in a graph. This interface is particularly useful for creating scenes that contain a very large number of elements interacting with each other. Popular industry node-based compositing software includes Fusion and Nuke.

During the 1980s and 1990s the general logic of computer representation—that is, representing everything as variables that can have different values—was systematically embedded throughout the interfaces of media design software. As a result, although a particular software application does not directly prescribe to its users what they can and cannot do, the structure of the interface strongly influences the designer's thinking. In the case of moving image design, the result of having a timeline interface with multiple channels all just waiting to be animated is that a designer usually does animate them. If previous constraints in animation technology—from the first optical toys in the early nineteenth century to the standard cel animation system in the twentieth century—resulted in an aesthetics of discrete and limited temporal changes, the interfaces of computer animation software quickly led to a new aesthetics: the continuous transformations of most (or all) visual elements appearing in a frame.

This change in animation aesthetics deriving from the interface design of animation software was paralleled by a change in another field—architecture. In the mid-1990s, when architects started to use software originally developed for computer animation and special effects (first Alias and Wavefront; later Maya and others), the idea of animated form entered architectural thinking as well. If 2D animation/compositing software such as After Effects enables an animator to change any parameter of a 2D object (a video clip, a 2D shapes, type, etc.) over time, 3D computer animation allows the same for any 3D shape. An animator can set up key frames manually and let a computer calculate how a shape changes over time. Alternatively, s/he can direct algorithms that will not only modify a shape over time but can also generate new ones. (3D computer animation tools to do this include particle systems, physical simulation, behavioral animation, artificial evolution, L-systems, etc.) Working with 3D animation software affected architectural imagination both metaphorically and literally. The shapes that started to appear in projects by young architects and architecture students in the second part of the 1990s looked as if they were in the process of being animated, captured as they were transforming from one state to another. The presentations of architectural projects and research begin to feature multiple variations of the same shape generated by varying parameters in software. Finally, in projects such as Gregg Lynn's New York Port Authority

Gateway (1995),[39] the paths of objects in an animation were literally turned into an architectural design. Using a particle system (a part of Wavefront animation software), which generates a cloud of points and moves them in space to satisfy a set of constraints, Lynn captured these movements and turned them into the curves making up his proposed building.

Equally crucial was the exposure of architects to the new generation of modeling tools in the commercial animation software of the 1990s. For two decades the main technique for 3D modeling was to represent an object as a collection of flat polygons. But by the mid-1990s, the faster processing speeds of computers and the increased size of computer memory made it practical to offer another technique on desktop workstations—spline-based modeling. This new technique for representing form pushed architectural thinking away from rectangular modernist geometry and toward the privileging of smooth and complex forms made from continuous curves. As a result, since the second part of the 1990s, the aesthetics of "blobs" has come to dominate the thinking of many architecture students, young architects, and even already well-established "star" architects such as Zaha Hadid, Eric Moss, and UN Studio.

But this was not the only consequence of the switch from the standard architectural tools and CAD software (such as AutoCAD) to animation/special effects software. Traditionally, architects created new projects on the basis of existing typology. A church, a private house, a railroad station all had their well-known types—the spatial templates determining the way space was to be organized. Similarly, when designing the details of a particular project, an architect would select from the various standard elements with well-known functions and forms: columns, doors, windows, etc.[40] In the twentieth century mass-produced housing only further embraced this logic, which eventually became encoded in the interfaces of CAD software.

But when in the early 1990s, Gregg Lynn, the firm Asymptote, Lars Spuybroek, and other young architects started to use 3D

---

[39] Gregg Lynn, *Animate Form* (Princeton Architectural Press, 1999), pp. 102–19.

[40] I am grateful to Lars Spuybroek, the principal of Nox, for explaining to me how the use of software for architectural design subverted traditional architectural thinking based on typologies.

software that had been created for other industries—computer animation, special effects, computer games, and industrial design—they found that this software came with none of the standard architectural templates or details. In addition, if CAD software for architects assumed that the basic building blocks of a structure are rectangular forms, 3D animation software came without such assumptions. Instead it offered splined curves and smooth surfaces and shapes constructed from these curves —which were appropriate for the creation of animated and game characters and industrial products. (In fact, splines were originally introduced into computer graphics in 1962 by Pierre Bézier for use in computer-aided car design.)

As a result, rather than being understood as a composition made up of template-driven standardized parts, a building could now be imagined as a single continuous curved form that can vary infinitely. It could also be imagined as a number of continuous forms interacting together. In either case, the shape of each of these forms was not determined by any kind of *a priori* typology.

(In retrospect, we can think of this highly productive "misuse" of 3D animation and modeling software by architects as another case of media hybridity—in particular, what I called the "crossover effect." In this case, it is a crossover between the conventions and the tools of one design field—character animation and special effects—and the ways of thinking and knowledge of another field, namely, architecture.)

Relating this discussion of architecture to the main subject of this chapter—production of moving images—we can see now that by the 1990s both fields were affected by computerization in a structurally similar way. In the case of twentieth century commercial animation, all temporal changes inside a frame were limited, discrete, and usually semantically driven—i.e., connected to the narrative. When an animated character moved, walked into a frame, turned his/her head, or extended his/her arm, this was used to advance the story.[41] After the switch to a software-based production process, moving images came to feature constant

[41] In the case of narrative animation produced in Russia, Eastern Europe and Japan, the visual changes in a sequence were not always driven by the development of a narrative and could serve other purposes—establishing a mood, representing the emotional state, or simply used aesthetically for its own sake.

changes on many visual dimensions that were no longer limited by semantics. As defined by numerous motion-graphics sequences and short films of the 2000s, contemporary temporal visual form constantly changes, pulsates, and mutates beyond the need to communicate meanings and narrative. (The films of Blake and Murata offer striking examples of this new aesthetics of a variable form; many other examples can easily be found by surfing websites that showcase works by motion graphics studios and individual designers.)

A parallel process took place in architectural design. The differentiations in a traditional architectural form were connected to the need to communicate meaning and/or to fulfill the architectural program. An opening in a wall was either a window or a door; a wall was a boundary between functionally different spaces. Thus, just as in animation, the changes in the form were limited and were driven by semantics. But today, the architectural form designed with modeling software can change its geometry across the whole design, and these changes no longer have to be justified by function.

The Yokohama International Port Terminal (2002), designed by Foreign Office Architects, illustrates very well the aesthetics of variable form in architecture. The building is a complex and continuous spatial volume without a single right angle and with no distinct boundaries that would break the form into parts or separate it from the ground plane. Visiting the building in December 2003, I spent four hours exploring the continuities between the exterior and the interior spaces and enjoying the constantly changing curvature of its surfaces. The building can be compared to a Mobius strip—except that it is much more complex, less symmetrical, and more unpredictable. It would be more appropriate to think of it as a whole set of such strips smoothly interlinked together.

To summarize this discussion of how the shift to software-based representations affected the modern language of form: all constants were substituted by variables whose values can change continuously. As a result, culture went through what we can call the *continuity turn*. Both the temporal visual form of motion graphics and design cinema and the spatial form of architecture entered the new universe of continuous change and transformation. (The fields of product design and space design were similarly affected.) Previously, such aesthetics of "total continuity" was imagined by only a few artists. For instance, in the 1950s, architect Frederick

Kiesler conceived a project titled *Continuous House* that, as the name implies, is a single continuously curving spatial form unconstrained by the usual divisions into rooms. But when architects started to work with 3D modeling and animation software in the 1990s, such thinking became commonplace. Similarly, the understanding of a moving image as a continuously changing visual form without any cuts, which previously could be found only in a small number of films made by experimental filmmakers throughout the twentieth century such as Fischinger's *Motion Painting* (1947), now became the norm.

Today, there are many successful short films lasting only a few minutes and small-scale building projects based on the aesthetics of continuity—i.e. a single continuously changing form, but the next challenge for both motion graphics and architecture is to discover ways to employ this aesthetics on a larger scale. How do you scale up the idea of a single continuously changing visual or spatial form, without any cuts (for films) or divisions into distinct parts (for architecture)?

In architecture, a number of architects have already begun to successfully address this challenge. Examples include already realized projects such as the Yokohama International Port Terminal, the Kunsthaus in Graz by Peter Cook (2004), and Ordos Museum by MAD Architects in Inner Mongolia, China (2012), as well as those under construction, such as Zaha Hadid's Performing Arts Centre on Saadiyat Island in Abu Dhabi, United Arab Emirates. (After the 2007 economic crisis, many ambitious building projects in Dubai and Eastern Europe were delayed or cancelled, but China and other countries continue to take risks and embrace the new architectural designs made from complex continuously changing curves.)

What about motion graphics? Blake has been one of the few artists to have systematically explored how hybrid visual language can work in longer pieces. *Sodium Fox* is 14 minutes; an earlier piece, *Mod Lang* (2001), is 16 minutes. The three films that make up *Winchester Trilogy* (2001–4) run for 21, 18, and 12 minutes. None of these films contain a single cut.

Sodium Fox and *Winchester Trilogy* use a variety of visual sources including photography, old film footage, drawings, animation, type, and computer imagery. All these media are woven together into a continuous flow. As I have already pointed out in relation

The Yokohama International Port Terminal. Foreign Office Architects, 2002.

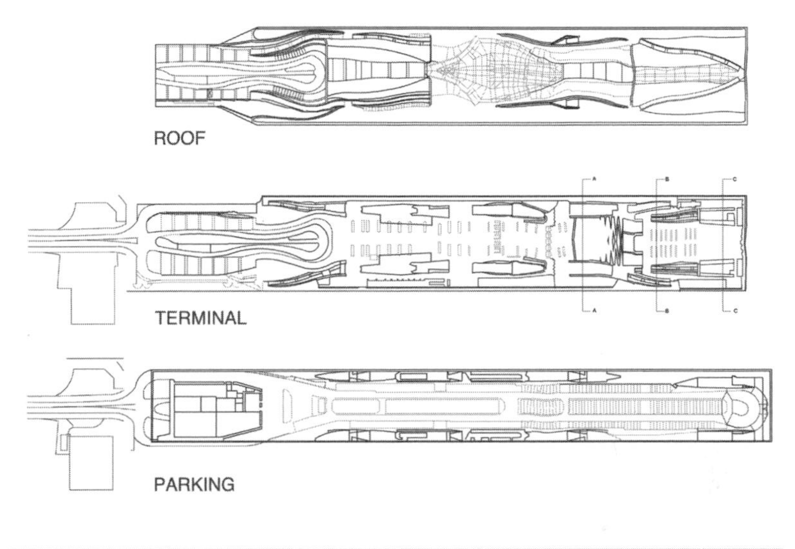
ROOF
TERMINAL
PARKING

to *Sodium Fox*, in contrast to shorter motion-graphics pieces with their frenzy of movement and animation, Blake's films contain very little animation in a traditional sense. Instead, various still or moving images gradually fade in on top of each other. So while each film moves through a vast terrain of different visuals—color and monochrome, completely abstract and figurative, ornamental and representational—it is impossible to divide the film into temporal units. In fact, even when I tried, I could not keep track of how the film got from one kind of image to a very different one just a couple of minutes later. And yet these changes were driven by some kind of logic, even if my brain could not compute it while I was watching each film.

The hypnotic continuity of these films can be partly explained by the fact that all visual sources in the films were manipulated via graphics software. In addition, many images were slightly blurred. As a result, regardless of the origin of the images, they all acquired a certain visual coherence. So although the films skillfully play on the visual and semantic differences between live-action footage, drawings, photographs with animated filters on top of them, and other media, these differences do not create juxtaposition or stylistic montage.[42] Instead, various media seem to peacefully coexist, occupying the same space. In other words, Blake's films seem to suggest that deep remix is not the only possible result of softwarization.

We have already discussed in detail Alan Kay's concept of a computer metamedium. According to Kay's proposal made in the 1970s, we should think of the digital computer as a metamedium containing all the different "already existing and non-yet-invented media."[43] What does this imply for the aesthetics of digital projects? In my view, it does not imply that the different media necessarily fuse together, or make up a new single hybrid, or result in "multimedia," "intermedia," "convergence," or a totalizing Gesamtskunstwerk. As I have argued, rather than collapsing into a single entity, different media (i.e., different techniques, data formats, data sources, and working methods) start interacting,

[42] In the "Compositing" chapter of *The Language of New Media*, I have defined "stylistic montage" as "juxtapositions of stylistically diverse images in different media."

[43] Kay and Goldberg, "Personal Dynamic Media."

producing a large number of hybrids, or new "media species." In other words, just as in biological evolution, media evolution in a software era leads to differentiation and increased diversity—more species rather than less.

In the world dominated by hybrids, Blake's films are rare in presenting us with relatively "pure" media appearances. We can either interpret this as the slowness of the art world, which is behind the evolutionary stage of professional media—or as a clever strategy by Blake to separate himself from the usual frenzy and over-stimulation of motion graphics. Or we can read his aesthetics as an implicit statement against the popular idea of "convergence." As demonstrated by Blake's films, while different media has become compatible, this does not mean that their distinct identities have collapsed. In *Sodium Fox* and *Winchester Trilogy*, the visual elements in different media maintain their defining characteristics and unique appearances.

Blake's films also expand our understanding of what the aesthetics of continuity can encompass. Different media elements are continuously added on top of each other, creating the experience of a continuous flow, which nevertheless preserves their differences. Danish artist Ann Lislegaard also belongs to the "continuity generation." A number of her films involve continuous navigation or an observation of imaginary architectural spaces. We may relate these films to the works of a number of twentieth-century painters and filmmakers who were concerned with similar spatial experiences: Giorgio de Chirico, Balthus, the Surrealists, Alain Resnais (*Last Year at Marienbad*), Andrei Tarkovsky (*Stalker*). However, the sensibility of Lislegaard's films is unmistakably that of the early twenty-first century. The spaces are not clashing together as in, for instance, *Last Year at Marienbad*, nor are they made uncanny by the introduction of figures and objects (a practice of René Magritte and other Surrealists). Instead, like her fellow artists Blake and Murata, Lislegaard presents us with forms that continuously change before our eyes. She offers us yet another version of the aesthetics of continuity made possible by software such as After Effects, which translates the general logic of computer representation—the substitution of all constants with variables—into concrete interfaces and tools.

The visual changes in Ann Lislegaard's *Crystal World (after J. G. Ballard)* (2006) happen right in front of us, and yet they are

Crystal World (after J. G. Ballard) *by Ann Lislegaard, 2006. Selected stills from a 3D animation.*

Crystal World (after J. G. Ballard) *by Ann Lislegaard, 2006. Installation photographs.*

practically impossible to track. Within the space of a minute, one space is completely transformed into something very different.

Crystal World creates its own hybrid aesthetics, one that combines photorealistic spaces, completely abstract forms, and a digitized photograph of plants. (Although I do not know the exact software Lislegaard's assistant used for this film, it is unmistakably some 3D computer animation package.) Since everything is rendered in gray scale, the differences between media are not loudly announced. And yet they are there. It is this kind of subtle and at the same time precisely formulated distinction between different media that gives this video its unique beauty. In contrast to twentieth-century montage, which created meaning and effect through dramatic juxtapositions of semantics, compositions, spaces, and different media, Lislegaard's aesthetics fits in with other minimalist cultural projects of the early twenty-first century. Today, the creators of minimal architecture and space design, web graphics, generative animations and interactives, ambient electronic music, and progressive fashions similarly assume that a user is intelligent enough to make out and enjoy subtle distinctions and continuous modulations.

Lislegaard's *Bellona (after Samuel R. Delany)* (2005) takes the aesthetics of continuity in a different direction. We are moving through and around what appears to be a single set of spaces. (Historically, such continuous movement through a 3D space has its roots in the early uses of 3D computer animation: first for flight simulators and later in architectural walk-throughs and first-person shooters.) While we pass through the same spaces many times, each time they are rendered in a different color scheme. The transparency and reflection levels also change. Lislegaard is playing a game with the viewer: while the overall structure of the film soon becomes clear, it is impossible to keep track of which space we are in at any given moment. We are never quite sure if we have already been there and it is now simply lit differently, or if it is a space that we have not yet visited.

Bellona can be read as an allegory of "variable form." In this case, variability is played out as seemingly endless color schemes and transparency settings. It does not matter how many times we have already seen the same space, it always can appear in a new way.

To show us our world and ourselves in a new way is, of course, one of the key goals of all modern art, regardless of the media.

By substituting all constants with variables, media software institutionalizes this desire. Now everything can always change and everything can be rendered in a new way. But, of course, simple changes in color or variations in a spatial form are not enough to create a new vision of the world. It takes talent to transform the possibilities offered by software into meaningful statements and original experiences. Lislegaard, Blake, and Murata—along with many other talented designers and artists working today—offer us distinct and original visions of our world in a state of continuous transformation and metamorphosis: visions that are fully appropriate for our time of rapid social, technological, and cultural change.

Amplification

Although the discussions in this chapter did not cover all the changes that took place during the Velvet Revolution, the magnitude of the transformations in moving image aesthetics and communication strategies should by now be clear. While we can name many social factors that all could have and probably did play some role—the rise of branding, experience economy, youth markets, and the Web as a global communication platform during the 1990s—I believe that these factors alone cannot account for the specific design and visual logics which we see today in media culture. Similarly, they cannot be explained by simply saying that contemporary consumer society requires constant innovation, constant novel aesthetics, and effects. This may be true—but why do we see these particular visual languages as opposed to others, and what is the logic that drives their evolution? I believe that to properly understand this, we need to carefully look at media creation, editing, and design software and its use in the production environment—which can range from one person with a laptop to a number of production companies around the world with thousands of people collaborating on the same large-scale project such as a feature film. In other words, we need to use the perspective of Software Studies.

The makers of software used in media production usually do not set out to create a revolution. On the contrary, software is created to fit into already existing production procedures, job roles, and familiar tasks. But software applications are like species within the common ecology—in this case, a shared environment of a digital computer. Once "released," they start interacting, mutating, and making hybrids. The Velvet Revolution can therefore be understood as the period of systematic hybridization between different software species originally designed to do work in different media. By 1993, designers had access to a number of programs which were already quite powerful but mostly incompatible: Illustrator for making vector-based drawings, Photoshop for editing of continuous tone images, Wavefront and Alias for 3D modeling and animation, After Effects for 2D animation, and so on. By the end of the 1990s, it became possible to use them in a single workflow. A designer could now combine operations and representational formats such as a bitmapped still image, an image sequence, a vector drawing, a 3D model and digital video specific to these programs within the same design. I believe that the hybrid visual language that we see today across "moving image" culture and media design in general is largely the outcome of this new production environment. While this language supports seemingly numerous variations as manifested in the particular media designs, its key aesthetics feature can be summed up in one idea: deep remixability of previously separate media languages.

As I have stressed more than once, the result of this hybridization is not simply a mechanical sum of the previously existing parts but new "species." This applies both to the visual language of particular designs and to the operations themselves. *When a pre-digital media operation is integrated into the overall software production environment, it often comes to function in a new way.* I would like to conclude by analyzing in detail how this process works in the case of a particular operation—in order to emphasize once again that media remixability is not simply about adding the content of different media, or adding together their techniques and languages. And since remix in contemporary culture is commonly understood as comprising these kinds of additions, we may want to use a different term to talk about the kinds of transformations the example below illustrates. I called this provisionally "deep remixability," but what is important is the idea and not a particular term. (So if you have a suggestion for a better one, send me an email.)

What does it mean when we see depth-of-field effects in motion graphics, films and television programs which use neither live action footage nor photorealistic 3D graphics, but have a more stylized look? Originally an artifact of lens-based recording, depth of field was simulated in software in the 1980s when the main goal of 3D computer graphics field was to create maximum "photorealism," i.e. synthetic scenes not distinguishable from live action cinematography. But once this technique became available, media designers gradually realized that it can be used regardless of how realistic or abstract the visual style is—as long as there is a suggestion of a 3D space. Typography moving in perspective through an empty space; drawn 2D characters positioned on different layers in a 3D space; a field of animated particles—any spatial composition can be put through the simulated depth of field.

The fact that this effect is simulated and removed from its original physical media means that a designer can manipulate it in a variety of ways. The parameters which define what part of the space is in focus can be independently animated, i.e. they can be set to change over time—because they are simply the numbers controlling the algorithm and not something built into the optics of a physical lens. So while simulated depth of field maintains the memory of the particular physical media (lens-based photo and film recording) from which it came, it became essentially a new technique which functions as a "character" in its own right. It has the fluidity and versatility not available previously. Its connection to the physical world is ambiguous at best. On the one hand, it only makes sense to use depth of field if you are constructing a 3D space, even if it is defined in a minimal way by using only a few or even a single depth cue, such as lines converging towards the vanishing point or foreshortening. On the other hand, the designer is now able to "draw" this effect in any way desirable. The axis controlling depth of field does not need to be perpendicular to the image plane, the area in focus can be anywhere in space, it can also quickly move around the space, etc.

In summary, when we remove depth of field from its original hardware home (photo and film cameras) and move it into software, we change where and how it can be used. We can still use it in its original context—that of lens-captured media. That is, we can apply it to 3D computer graphics elements in order to make them compatible with video or film captured via lens. But we can also now use it with many other media, for purely artistic effect.

And we can use it in many ways that were not even conceivable when it was part of media hardware.

Following the Velvet Revolution, the aesthetic charge of many media designs is often derived from "simpler" remix operations—juxtaposing different media in what can be called "media montage." However, for me the essence of this Revolution is the more fundamental "deep remixability," illustrated by this example of how depth of field was greatly amplified when it was simulated in software.

Computerization virtualized practically all media creation and modification techniques, "extracting" them from their particular physical media and turning them into algorithms. This means that in most cases, *we will no longer find any of the pre-digital techniques in their pure original state*. This is something I have already discussed in general when we looked at the first stage in cultural software history, i.e. the 1960s and 1970s, examining Sutherland's work on the first interactive graphical editor Sketchpad, Nelson's concepts of hypertext and hypermedia, and Kay's discussions of an electronic book ("It [an electronic book] need not be treated as a simulated paper book since this is a new medium with new properties"). We have now seen how this general idea articulated already in the early 1960s made its way into the details of the interfaces and tools of applications for media design which eventually replaced most of traditional tools: After Effects (which we analyzed in detail), Illustrator, Photoshop, Flash, Final Cut, etc. So what is true for depth of field effect is also true for most other tools offered by media design applications.

What was a set of theoretical concepts implemented in a small number of custom software systems accessible mostly to their own creators in the 1960s and 1970s (such as Sketchpad or the Xerox PARC workstation) later became a universal production environment used today throughout all areas of the culture industry. The ongoing interactions between the ideas coming from the software industry and the desires of the users of their tools (media designers, graphic designers, film editors, and so on)—led to the further evolution of software—for instance, the emergence of an new category of "digital asset management" systems around the early 2000s, or the concept of "production pipeline" which became important in the middle of this decade. In this chapter I have described only one among many directions of the evolution of software applications, their tools, and media formats. As we saw, the result of this trend was the emergence of a fundamentally new type of aesthetics which today dominates visual and media culture.

CONCLUSION

Software, hardware, and social media

In 1977 Alan Kay and Adele Goldberg imagined that a computer would become a "metamedium" that would contain "a wide range of already-existing and not-yet-invented media." Exactly as they predicted, computers have been used to invent a number of *new types of media that are not simulations of prior physical media*. The examples include navigable three-dimensional spaces constructed with computer graphics, media databases, or "sim" video games such as SimCity, The Sims, and Civilization. And, of course, the Internet in particular has been a very productive host for inventing new types of communication and collaboration: email, forums, blogs, microbloging (e.g. Twitter), wiki, RSS, social networks, etc.

At the same time, the computer metamedium was also evolving in a second direction. While it indeed came to contain "a wide range of already-existing" media, I have argued that once each of these media was simulated in a computer their identities changed. Sutherland, Engelbart, Nelson, Kay, and other inventors of computer metamediums understood that *the simulations of previously existing physical or electronic media could add many new properties to these media*. As Kay and Goldberg wrote in their article, a simulated medium can become a "new medium with new properties."

A computer "breathes new life" into the physical and electronic media it simulates. Media can become "dynamic" (to use another of Kay's terms which he preferred to "interactive"). They can also be "intelligent": think of Sketchpad, which could automatically "clean up" the sketches made by designers by satisfying constraints such as parallelism. Further, they can become collectively sharable and collectively "editable"—think of large-scale social software projects such as Wikipedia and OpenStreetMap. Media objects such as pictures, sound, video, text, and code can leap from machine to machine in a truly magical fashion: from a mobile phone to a media player, then to a memory card, a laptop, a netbook, the Web, and so on and so on. Imagine that you live in the sixteenth century and you are told that you can order an image in a painting to travel by itself and appear in another painting in another country, or that a text in one book can lift itself and replace a text in another book? And yet this is exactly what many of us are doing every day without even thinking how magical this is.

I think that the scale, diversity and radicalism of these "additions"—those already invented and many more yet to be invented in the future—is of such magnitude that exploring what can be created with them will occupy us for a long time. And this is one of the reasons why the "digital revolution" is different from all previous techno-cultural revolutions. The ability to simulate not simply one or two, but most media in a computer—combined with computer abilities to control processes in real-time, calculate, transform inputs, test what/if scenarios, and send information over networks—opens a practically unbounded space of creative possibilities.

As we followed the evolution of the computer medium from the first stage of invention and experimentation (1960s–1970s) to the second stage of commercialization and wide adoption (1980s–1990s), we discovered the third direction of the computer metamedium evolution—*hybridization*. Translated into software, different types of media started acting like species within a common ecology—in this case, a shared software environment. Once they were "released" into this environment they begin interacting, mutating, and making hybrids. Both the simulated and new media types—text, hypertext, still photographs, vector graphics, digital video, 2D animation, 3D models and animation, navigable 3D spaces, maps, location information, messages and scripts—became building blocks for many new media combinations. As my examples illustrate, such hybrids can be found at different scales—from large-scale software systems such as Google Earth to the single images and short motion graphics created by individual designers.

The rise of social media and social networking on the web in the middle of the 2000s, their expansion to mobile platforms in the next few years, and the development of apps markets for these platforms lead to new types of hybrids. The functional elements of social and mobile media—search, rating, wall posting, subscription, text messaging, instant messaging, email, voice calling, video calling, etc.—form their own media ecosystem. Like the ecosystem of techniques for media creating, editing and navigation realized in professional media software during the 1990s, this new ecosystem enables further interactions of its elements. These elements are combined in a variety of ways on different platforms and in different social media apps. New "features" of social and mobile software enter this ecosystem, and expand the pool of its elements.

The technologies behind these elements, such as multi-tier web applications and server-side scripting, make creating new combinations and new elements particularly easy in comparison to desktop applications.[1] While new versions of desktop applications such as Photoshop, InDesign or Maya that may contain new techniques are released infrequently, online applications such as Facebook, YouTube or Google Search can be updated by their companies very frequently (both Google and Facebook update their code daily) and new features can be added at any time.

The competition between leading social network services has been one engine of new elements as well as variations on existing elements. To give two examples from 2011, Google+ introduced the "Circles" feature to enable users to organize other people using the service into groups and to control what they post and share with each group;[2] Facebook introduced a Subscribe button to allow its users to follow public updates from particular users they like.[3] In the same year, the new social photo sharing service Pinterest rose meteorically, due to its new features such as a flexible layout of photos in contrast to the fixed grid layout of Facebook and Google+.

---

[1] For a good explanation of how web applications work technically, see http://en.wikipedia.org/wiki/Web_application

[2] Vic Gundotra, "Introducing the Google+ project: Real-life sharing, rethought for the web," googleblog.blogspot.com, June 28, 2011, http://googleblog.blogspot.com/2011/06/introducing-google-project-real-life.html

[3] Meghan Peters, "Facebook Subscribe Button: What It Means for Each Type of User," mashable.com, September 15, 2011, http://mashable.com/2011/09/15/facebook-subscribe-users/

While this book focused on the evolution of software for media creation and editing, this evolution is intrinsically connected with the parallel development of *hardware*, including desktops, and later laptops and mobile platforms, networks, servers, render farms, video cards, displays, and so on. More high resolution displays, larger and cheaper storage, faster networks, easier connectivity between devices for media capture, storage, editing, distribution, and playback—all these developments automatically extend the capacities of computational media, changing what can be imagined and designed. For example, a computer display with 35,840 x 8,000 pixel resolution (e.g. the HIPerSpace super-visualization computer that my lab has been fortunate to use since 2008) is not only quantitatively but also qualitatively different from a display which only has 1024 x 768 pixels (the desktop standard in the 1990s).[4] Similarly, the experience of using the Internet via broadband connection is qualitatively different from a modem to dial-in via analog telephone line (also the standard in the 1990s).

When Kay's colleagues implemented various media editors on their "interim Dynabook" in the first part of the 1970s, most of this software could not compete with their physical and electronic equivalent tools. I remember my experience working on a first generation Macintosh in 1984; it could only show sixteen levels of grey on a 512 x 384 pixel screen measuring 9 inches. Obviously this was not sufficient for me to immediately drop my oil brushes and paints and switch to computers. So, in some sense, the first period of media computerization—between the completion of Sketchpad in 1963 and the release of PageMaker in 1985—was theoretical. During this period, the conceptual principles and the key algorithms necessary for detailed simulation of physical media were developed, before the sufficiently cheap hardware was available. For instance, during the 1960s many computer scientists learned about Sketchpad by reading Sutherland's PhD thesis since the machine on which it run—the TX-2 computer—existed only at MIT. (This is another interesting characteristic of computer media revolution—it was theorized in detail before it occurred in practice.)

---

[4] Photographs of my lab's work with the HIPerSpace super-visualization computer are available at http://lab.softwarestudies.com/2008/12/cultural-analytics-hiperspace-and.html. The HIPerSpace computer is described at http://vis.ucsd.edu/mediawiki/index.php/Research_Projects:_HIPerSpace

But during the second part of the 1990s PC hardware become advanced enough to run simulations of most media at a sufficient fidelity that was comparable with the professional standards already in place. This included software for graphic design, CAD, 3D modeling, 2D and 3D animation, print layout, digital photo manipulation, and audio and video non-linear editing. In many cases, these applications were now responding to users' actions sufficiently fast to approach the level of interactivity conceptualized by Kay and Goldberg in 1977 (although 3D animation and video compositing even today may require long rendering times). As a result, simulated media became truly useful, and also accessible, to large numbers of people outside of computer laboratories and large media companies. Within half a decade, most culture professionals abandoned physical media for their simulated equivalents.

When I visited well-known electronic musician, writer and artist DJ Spooky in his Tribeca apartment in New York in January 2005, I did not find any musical instruments, traditional or electronic. The only "instrument" Paul Miller (aka DJ Spooky That Subliminal Kid) owned was his 15-inch PowerBook laptop, made by Apple. This was his "Dynabook": a "self-contained knowledge manipulator in a portable package the size and shape of an ordinary notebook."[5] Although this "Dynabook" did not have Smalltalk, it ran another programming environment which was powerful, fast and allowed for visual programming—MAX, the language of choice worldwide for tens of thousands of electronic musicians, VJs, dancers, theatre performers and others working with different forms of real-time performance.

While the evolution of hardware enabled dissemination of media software to professional communities in the 1990s, in the 2000s this evolution also enabled the new stage in software design and use—social media (2004–). The few companies which dominated the field of professional media applications (Adobe, Apple, Autodesk) were joined by a multitude of new companies and countless start-ups focused on developing tools and services for the Web and mobile platforms.

The new software categories include social networking (e.g. Facebook), micro-content services (e.g. Twitter), media sharing

---

[5] Kay and Goldberg, "Personal Dynamic Media," p. 394.

websites (YouTube, Vimeo, Picasa, Flickr, etc.); consumer-level software for media organization and light editing (e.g., iPhoto); blog editors (Blogger, WordPress); and many others. Keep in mind that software—especially web and mobile applications and services designed for consumers—continuously evolves, so some of the categories above, their popularity, and the features of particular applications and social network services may change by the time you are are reading this. One graphic example is the shift in the identity of Facebook. During 2007, it moved from being yet another social media application that was competing with MySpace to a "social OS" that aimed to combine the functionality of previously different applications in one place—replacing, for instance, stand-alone email software for many users.)

None of the software apps and websites of the "social media era" function in isolation. Instead, they participate in the larger ecology, which includes search engines, recommendation engines, blogging systems, RSS feeds, and other web technologies; inexpensive consumer electronic devices for capturing and accessing media (digital cameras, mobile phones, music players, video players, digital photo frames, internet enabled TVs); and technologies that enable transfer of media between devices, people, and the web (storage devices, wireless technologies such as Wi-Fi and WiMax, and communication standards such as USB and 4G). Without this ecology most web services and mobile apps would not be possible. Therefore, this ecology needs to be taken into account in any discussion of social networks and their software—as well as consumer-level content access and media development software designed to work with web-based media sharing sites. And while the particular elements and their relationship in this ecology are likely to change over time—for instance, almost all media content may eventually be available over computer networks; communication between devices may similarly become fully transparent; and the very rigid physical separation between people, devices they control, and "non-smart" passive space may become blurred—the very idea of a technological ecology consisting of a number of interacting parts which include software is not unlikely to go away any time soon. Thus, if one day I were to write a detailed account of social media, I would need to discuss consumer electronics, network architectures and protocols and other elements of this ecology as much as the social software itself.

Media after software

Any summary of a 100,000-word book of theoretical arguments can't cover all important points. However, I will risk this because I know that even an incomplete summary will still be useful for the readers. Here are some of the proposals developed in this book about the experiences and meanings of "media" for contemporary designers who create it using software applications, and for the users of interactive media applications and services:

1. The computer is not a new "medium"—it is the first "metamedium": a combination of existing, new, and yet to be invented media. (This is the argument of Kay, which I take as my starting point.)
2. A "medium" (as it exists in software) is a combination of particular techniques for generation, editing and accessing content. (I use the generic term "access" as an alias to the longer list of terms—navigating, browsing, viewing, listening, reading, and interacting.) Softwarization virtualizes already existing techniques and adds many new ones. All these techniques together from the "computer metamedium." Any single "medium" uses a subset of these.

   New techniques and their variations are constantly being developed which changes the identity of each medium that uses them. For the users of popular commercial media software, a medium changes with each software release.
3. "What we identify by a conceptual inertia as "properties" of different mediums are actually *the properties of media software* —their interfaces, the tools, and the techniques they make possible for accessing, navigating, creating, modifying, publishing and sharing media documents."
4. Following the same logic, "properties" of any media object are no longer fully defined by the content and formats of the files storing the information. They now also depend on the software used to access this object. For example, depending on whether the same image is accessed via a default media viewer, a consumer app for media access and

editing, or professional editing software such as Photoshop, its "properties" change significantly.

5 The techniques that make up the computer metamedium can be divided into two categories. General-purpose (i.e. "media-independent") techniques are implemented to work in the similar way on all media types (for example: select, copy, search, filter, etc.) Media-specific techniques can only work on particular data structures (for example, you can increase amplitude of a sound track or reduce the number of vertices in a 3D shape, but not vice versa). Each software "medium" combines some media-independent and some media-specific techniques.

6 The idea of a data structure leads us to an alternative definition of a software medium. "As defined by application software and experienced by users, a 'medium' is a pairing of a particular data structure and the algorithms for creation, editing and viewing the content stored in this structure."

From this perspective, each of the categories of media development software can be said to define its own "medium"—because the programs offered in each of these categories typically (but not always) share the same fundamental data structure. The examples that fit this are vector graphics editors, raster graphics editors, 2D animation and motion graphics software, 3D computer graphics software, sound editors, text processors, and HTML editors.

7 Following the first stage of the computer metamedium invention, we enter the next stage of media "hybridity" and "deep remixability." "The unique properties and techniques of different media became software elements that can be combined together in previously impossible ways." "Both the simulated and new media types—text, hypertext, still photographs, digital video, 2D animation, 3D animation, navigable 3D spaces, maps, location information—came to function as building blocks for many new media combinations."

This condition is not the simple consequence of the universal digital code used for all media types. Instead, it is the result of the gradual development of interoperability technologies including standard media file formats, import/export functions in applications, and network protocols.

8 The previous formulations lead us to view contemporary media development using a model of biological evolution and its concept of massive numbers of species that share common traits—away from the modern model of a small number of very different mediums with their unique languages. Instead of trying to place any particular project, app, or web service in some category selected from a small number, we can instead view it as a combination of the techniques selected from a very large pool. Some of these combinations occur more often; others may only occur once. The successful combinations become popular, leading to similar projects; and some may become design patterns used in numerous applications.

Software epistemology

One of the key ideas developed in this book is that the computer metamedium is characterized by "permanent extendibility." New algorithms and techniques that work with common media data types and file formats can be invented at any time by anyone with the right skills. These inventions can be distributed instantly over the web, without a need for the large resources that were required in the twentieth century to introduce a new commercial media device with new functions or a new media format. Use of open source and free software licenses and web-based hosting services repositories such as GitHub encourages people to collectively expand existing software tools, which often leads to their rapid evolution.

The permanent extendibility of the computer metamedium has important consequences not only for how we create and interact with media, but also for the *techniques of knowledge in a "computerized society"* ("Knowledge in Computerized Societies" is the subtitle of the opening section of the celebrated 1979 book *The Postmodern Condition: A Report on Knowledge* by Jean-François Lyotard).

Turning everything into data, and using algorithms to analyze it changes what it means to know something. It creates new strategies that together make up *software epistemology*. Epistemology is a branch of philosophy that asks questions such as what is knowledge, how it can be acquired, and to what extent a subject can be known. Digital code, data visualization, GIS, information retrieval, machine learning techniques, constantly increasing speed of processors and decreasing cost of storage, big data analytics technologies, social media, and other parts of the modern techno-social universe introduce new ways of acquiring knowledge, and in the process redefine what knowledge is.

For instance, it is always possible to invent new algorithms (or new ways to scale existing algorithms to analyze big data faster) that can analyze the already existing data in ways the previous algorithms could not. As a result, you can extract additional patterns and generate new information from the older, already analyzed data.

Algorithms and software applications that analyze images and video provide particularly striking examples of this capacity to generate additional information from the data years or even decades after it was recorded.

In *Blowup*, a 1966 film directed by Michelangelo Antonioni, the photographer who takes pictures of a couple kissing in the park uses the strategy of printing progressive enlargements of one area of the photograph, until a grainy close-up reveals a body in the grass and a man with a gun hiding in the trees.

During the time that this film was being created and unknown to its director, computer science researchers were already developing the new field of digital image processing, including algorithms for image enhancement such as sharpening of edges, increasing contrast, and reducing noise and blur. The early articles in the field show the blurry photographs taken by surveillance planes, which were sharpened by the algorithms. As I already explained earlier, today many of these techniques are built into all image manipulation software such as Photoshop, as well as in the firmware of digital cameras. They became essential to both consumer photography and commercial visual media, as every published photograph in mass media first goes through some software adjustments.

Contemporary DSLR models and high-end compact digital cameras can record images both in JPEG and Raw formats. With JPEG format, an image is compressed, which limits the possibilities for later extraction of additional information using software.

Raw format stores the unprocessed data from the camera's image sensor. The use of this format assumes that a photographer will later manipulate the photo in software to get the best results from the millions of pixels recorded by the camera. In his guide for the use of the two formats by photographers, William Porter explains: "Working with a Raw file, you'll have a greater chance of recovering highlights in overexposed clouds, rescuing detail from areas of shadow or darkness, and correcting white balance problems. You'll be able to minimize your images' noise and maximize their sharpness, with much greater finesse."[6]

This means that new software with better algorithms can generate new information from a photo in Raw format captured years earlier. (See the examples from Porter's article available online for a dramatic demonstration of this.[7]) In another example of software epistemology, in the demo presented at one SIGGRAPH annual conferences, a few film shots of 1940s' Hollywood film were manipulated in software to re-generate the same shots from a different point of view.

In visual effects production today, one of the most widely used operations is the algorithmic extraction of the position of the video camera that was used to capture a shot. This operation is called "motion tracking" and it exemplifies how information that is not directly available in the data can be inferred by algorithms. (The extracted camera position is used to insert computer graphics into the live action footage in the correct perspective.)

Another important type of software epistemology is *data fusion—using data from different sources to create new knowledge* that is not explicitly contained in any of them. For example, using the web sources, it is possible to create a comprehensive description of an individual by combining pieces of information from his/her various social media profiles and making deductions from them.

Combining separate media sources can also give additional meanings to each of the sources. Consider the technique of the automatic stitching of a number of separate photos into a single panorama, available in most digital cameras. Strictly speaking, the

---

[6] William Porter, "Raw vs. JPEG: Which should you shoot and when?", techhive.com (September 19, 2012).

[7] *Ibid.*

underlying algorithms do not add any new information to each of the images (i.e., their pixels are not modified). But since each image now is a part of the larger panorama, its meaning for a human observer changes.

The abilities to generate new information from the old data, fuse separate information sources together, and create new knowledge from old analog sources are just some techniques of software epistemology. In my future publications I am hoping to gradually describe other techniques—as I am teaching myself data mining and other algorithmic knowledge techniques of knowledge commonly used by contemporary software societies.

In the beginning of the book I asked: *what is media after software*? If artistic mediums were traditionally defined by the techniques and representational capacities of particular tools and machines (brushes, ink, paper, musical instruments, a printing press, a photo camera, a film camera, video cameras and editing equipment), what happens to this concept after most of these tools and techniques are simulated in a single software environment? In other words, what is a "medium" as defined by software applications used to create, edit, distribute, and access it?

While "media effects," "media representation," "media industry," "media theory" and "media history" have been extensively discussed in large number of books and articles in a number of academic disciplines, this literature does not usually include analysis of software tools and platforms. In contrast, the vast universe of "how to" books, instructional videos and tutorials contains very little theory, because the goal of all these publications is practical instruction. (My search for "Photoshop" in amazon.com under books on August 12, 2013 returned 9,405 publications; the search for "After Effects" returned 1,201 results, and the search for "3ds Max" returned 1,972 results.) The aim of my book was to help bridge the gap between these two separate universes of theory and practice.

Following the question of what it means to create media with software took us on a long journey through a few decades. We did find some possible answers that I hope you have found interesting and provocative. But of course, since both this book and all my writing are directed first of all to media practitioners—professionals and students creating new software applications and tools, graphic designs, web designs, motion graphics, animations, films, space designs, architecture, objects, devices, and digital art—you can do more than simply agree or disagree with my analysis. By inventing new techniques, or through the innovative application of existing techniques—and by finding new ways to represent the world, the human being, and the data, and new ways for people to connect, share, and collaborate—you can expand the boundaries of "media after software."

INDEX

A to D *see* analog to digital
Abel, Robert 288
Abyss, The 21, 294
ACM Digital Library 38–9
Acrobat 38, 269
  8.0 188
  designers 189
Acrobat Reader 195, 225
Acrobat User Interface 188
ActionScript 16
Adobe 27, 46, 93, 113, 147, 211, 283, 302, 333
  website 304
Adobe Acrobat *see* Acrobat
Adobe Bridge 144
Adobe Creative Suite 302
Adobe Photoshop Touch 26
Adobe Premiere Pro 304
Advanced Research Projects Agency Network 183–4
aerial imagery 134
aesthetic theory 149, 150
aesthetics 4, 45, 120, 121–2, 182, 236, 244, 252, 258, 259, 277, 297, 303, 311, 318, 319, 322, 323, 324, 325
  Baroque 215
  hybridity 254–67
  psychedelic 298
After Effects 2, 24, 39, 41, 44, 45, 46, 49, 58, 75, 84, 90, 124, 144, 179, 205, 246, 247, 248, 249, 255, 261, 278, 282–9, 291, 293, 296, 300, 302, 305, 306, 307, 310, 311, 319, 324, 325
  4.0 47
After the Great Divide 250
Agre, Phil 12
Ajax techniques 218
Aldus 283
Aldus Pagemaker 41
Alexander, Amy 13–14
algorithms 114, 128, 136, 137, 138–9, 146, 179, 181, 182, 184, 197, 199–204, 215, 216, 219, 220, 222, 270, 325, 337, 338, 340
Algorithms Plus Data Structure Equals Programs 207
Alias 41, 44, 148, 290, 311, 324
AliasWavefront 148
Alice in Wonderland 259
Alien 294
Alsace 185, 187, 191, 196, 197, 237
Amazon 7, 9, 35, 92, 190, 191
Amazon Kindle 109
amplification 323–7
analog to digital 153
Anderson, Peter 299, 303
Android 7, 24, 26, 29, 182, 228
animation 29, 129, 145, 170
Anime Music Video 46
Antonioni, Michelangelo 338
Apeloig, Philippe 300, 303

Aperture 24, 124, 144, 225
API 190, 191, 237
App Store 108
Apple 6, 7, 39, 47, 57, 84, 85, 93, 105, 107, 108, 109, 147, 164, 225, 333
statistics 108
Apple IIe 20
AppleScript 211
apps 9, 17, 24, 25, 31, 47, 50, 108, 124, 148
architecture 41
Architecture Machine Group 161
Arns, Inke 14
ARPANET *see* Advanced Research Projects Agency Network
art 162
Art+Com 170, 187, 192
Artstor 228
artworks 139
ASCII text 209
Aspen Movie Map 80, 161, 164, 165, 176, 181
Asympote 312
Atari 84
Atkinson, Bill 105
authoring 163
AutoCAD 41, 44, 75, 312
Autodesk 27, 85, 283, 333
automation 128, 129
Avatar 259
Avid 59, 283

Bach, J. S. 118
Balthus 319
bandwidth 134
Bangalore 32
Barthes, Roland 81
Bass, Saul 40, 46, 279, 288
BBC
Your Painting project 227, 228
Beach Boys 145
Beatles 145
Bell, Alexander Graham 155
Bell, Daniel 100
Bell Laboratories 57, 85
Bellona 322
Berners-Lee, Tim 21, 161
Bézier, Pierre 313
Bilal, Enki 259
Bing Maps 143, 196
biological evolution 167–8, 177, 233
theories 168
bitmap image 209
bit-mapped display 99
Blake, Jeremy 258, 260–1, 307, 314, 315, 319, 323
Blender 24, 39, 203
Blogger 2, 24, 27, 47, 206, 334
blogs 1
Blowup 338
Blue Brain Project 236
Bogost, Ian 15, 42
Bolter, Jay 15, 58, 59, 61
books 78
bootstrapping 83
Bordwell, David 276
Borges, Jorge 11
brands, global 6
Breathless 81
browsers 35, 50, 150, 193
Bruges, Jason 164
Brunelleschi, Filippo 40
Bruner, Jerome 98, 233
theory of multiple mentalities 98, 100
Burton, Tim 259
Bush, Vannevar 11, 63, 83
Memex 73

C++ 238
CAD *see* Computer Aided Design
Cameron, James 21, 58, 259
Cars 280

Cartesian space 294
Castells, Manuel 15, 16
CD-ROMs 166
cel animation 145
CGI 259, 290, 291, 295
chat 2
children 98, 102, 104
Chiroco, Giorgio de 319
Chrome 2, 24
Chun, Wendy 15
cinema 170, 236
cinematography 193, 199, 274
CinePaint 206, 247
Cisco Systems 7
Civilization 329
CNN
  website 6
code studies 42
Codecademy 17
coding 133
cognition 123
cognitive psychology 232
Coldplay 257
color reproduction 299
Combustion 246
Coming of Post-Industrial Society, The 100
commands 114, 212, 225, 297
Common 257
communication 205
communication studies 34, 228
communication theory 35
compositing 277–82
Computational Culture 12
computational media 91, 93, 94, 96, 97, 100, 135
Computer Aided Design (CAD) 93, 230, 283, 312, 313, 333
computer drawing 93
computer graphics 89, 118
Computer History museum 40
computer industry 55
computer metamedium 45
computer painting 93
computer programming 208, 220, 223, 239
computer programs 34
computer science 10
computer scientists 31, 84, 91, 156
constructivism 121, 122
contrast 299
Convert 257, 268
Cook, Peter 315
Cox, Geoff 14
crafts 2
Cramer, Florian 14, 15
creativity 97
Crystal World 319, 322
Cubism 83
cultural anthropologists 32
cultural heritage
  digitization 226
cultural memory 183
cultural studies 205
cursor 128
cut and paste 114, 122
cyberculture 75
cyberspace 75
cyborgs 75

D to A *see* digital to analog
dada sonification 115
Darwin, Charles 176–7, 239
  On the Origin of Species 176
Darwinism
  literary 33
data manipulation 110–12
data structures 197, 207, 201, 220
databases 38–9, 143
Datapoint Corporation 85
DC Comics 180
De Stijl 82
  aesthetics 121
de.licio.us 84

deep remix 45, 270, 318
deep remixability 305, 324, 325
Derrida, Jacques 81
design
  visual 45
design studies 205
designers 93
desktops 60
Diesel 164
Digg 84
digital art 139, 166
Digital Art 162
digital compositing 144–7
digital computers 59, 135
digital culture 25
Digital Darkroom 203
Digital Effects 250
digital imaging 201
digital libraries 38–9
digital media 59, 132, 152, 155, 219
  scholars 81
digital photography 61–2
digital to analog 153
digital video recorder (DVR) 35
Director 21, 44
Disney 85, 266
DJ Spooky 333
DJ/VJ/live cinema performances 65
Dourish, Paul 12
Dow Jones Industrial Average 6–7
Dreamweaver 24, 39, 124, 165, 206, 247, 302
DSLR 338
Duff, Thomas 145, 146
Duncan, Isadora 40
DVD players 153
DVR *see* digital video recorder
Dynabook 64–91, 109, 231, 333

Earl King 80
Earth Viewer 192
eBay 7, 9
Edison, Thomas 154, 155, 156
editing 163
Eisenstein, Sergei 40
electronic devices 108
electronic literature 42
electronics industries 223
email 2
encoding 154
Engelbart, Douglas 5, 13, 40, 56, 60, 63, 71, 72, 73, 75, 78, 83, 91, 92, 93, 96, 97, 102, 104, 135, 162, 184, 225, 329
epistemology 337–41
EPS 30
Epstein, Jean 119
Equalize 132
Ethernet 57
Europeana 227, 228
evolution 45, 168
evolutionary biology 237, 239
evolutionary theory 177
Excel 30, 220
exegesis 78
Expanded Cinema 92
export 296–307
Expressionism 83
Expressive Processing 103
Extrude 130

Facebook 1, 2, 6, 9, 24, 27, 30, 39, 47, 84, 124, 148, 149, 331
fair trade certification 37
Fall Joint Computer Conference 72
Farm Security Administration 227
feature films 179
Ferro, Pablo 46, 254, 279, 298
Fessenden, Reginald 155
FFmpeg 211
Field Studies 185

file formats 215–19
files 135
film and TV studies 205
Final Cut 2, 24, 32, 39, 75, 118, 124, 144, 179, 206, 247, 283, 325
Final Cut Pro 47
findability 115, 119, 122, 124, 151
fine art 2
Finlay, Karen 164
Firefox 2, 24, 47, 92
Fischinger, Oskar 266, 286–7, 298, 315
Flame 39, 206, 246, 252, 294
Flash 31, 84, 179, 206, 246, 300, 302, 325
Flickr 30, 38, 47, 84, 92, 148, 150, 184, 191, 196–7, 227, 334
Flickrvision 3D 189
folders 29, 101, 135
Ford 7
Foreign Office Architects 314
Foucault, Michel 81
4G 334
Foursquare 38
Frame Imagery 279
Frantz, Matt 248
Freehand 58
Freud, Sigmund
  Interpretation of Dreams 169
Fry, Ben 105
Fujihata, Masaki 185, 187
Fuller, Matthew 11, 14, 15
functionality 221
functions 137, 220, 221
Fusion 247
Futurism 82, 83
futurists 236

Galloway, Alex 15
game platforms 42
game studies 205
GarageBand 145
Gardner, Howard 233
General Electric 7
General Motors 7
General Perspective Projection 36
Geocommons 38
Geographical Information Systems (GIS) 45–6, 338
Gesamtkunstwerk 120, 318
Giedion, Sigfried 5
Gimp 39, 50, 205, 247
GIS *see* Geographical Information Systems
Glass, Philip 118
global culture industry 27
global information society 8
globalization 8, 85
Gmail 24
GMC Denali 257
Go 268, 271, 293
Godard, Jean-Luc 279
Goldberg, Adele 64, 70, 97, 99, 100, 102, 104, 105, 110, 123, 140, 141, 150, 161, 163, 176, 180–81, 329, 333
Goodman, Nelson 65
  Languages of Art 65
Google 1, 6, 7, 9, 23, 24, 27, 39, 148, 149, 151, 190, 191, 198, 226, 287
  algorithms 71
  SketchUp 203
Google Analytics 30
Google Books 123
Google Docs 25, 220
Google Earth 2, 36, 37, 38, 45, 48, 84, 143, 148, 163, 165, 171, 182, 187, 192, 193, 195, 196, 204, 218, 234, 237, 277
  5.0 192, 195

6.0 193
GitHub 149, 330, 337
Street View 36, 164, 195
Google Image Search 114
Google Maps 16, 24, 37, 38, 47, 92, 143, 184, 190, 193
Google Play 17, 26, 35
Google Scholar 147
Google Search 47, 331
Google Trends 50, 118
Google+ 331
Goriunova, Olga 14
GPS 182, 184, 185, 192
graphic design 170
Graphic Design for the 21st Century 299
Graphical User Interface (GUI) 20, 21, 55, 57, 62, 97, 99, 100, 101, 102, 108, 148, 188, 189, 211, 212, 221, 222, 233, 237, 283
applications 109
graphics card 91
Graz, Kunsthaus 315
Greenberg, Clement 120, 236
Greenberg, Richard 288
Greenberg, Robert 288
Greenpeace data 37
Griffith, D. W. 40
Grusin, Richard 58, 59, 61
GUI *see* Graphical User Interface
Gutenberg, Johannes 40
Gutenberg Bible 94

hackers 91, 93
Hadid, Zaha 248, 300, 303, 312, 315
Hall, Stuart 35
Hamilton, Richard 251
hard drives 32
hardware 92, 96, 332, 333
Harrell, Fox 12
Harry 251
Hayles, Katherine 14, 15–16
HCI
designers 100
experts 100
HD resolution 154
HDTV 302
Hearfield, John 278
Hewlett-Packard 84
High-pass 132
HIPerSpace 332
Hiroshima Mon Amour 81
Hitchcock, Alfred 279
Hoch, Hannah 278
Hockney, David 251
hosting websites 28
HTML 42, 84, 170, 179, 201, 209, 218
markup 31
HTML5 80
Hugo 259
Human Connectome Project 236
human perception 123
humanists 32
Huyssen, Andreas 250
hybrid media 166, 171
hybrid visual language 258
hybridity 161–76, 184–95, 204, 257, 267
aesthetics 254–67
hybridization 176, 200, 247, 324, 330
HyperCard 21, 84, 105
hyperfilm 79–80
hyperlinking 71, 151, 178–9
hypermedia 71, 79, 161, 274, 325
hypermodernity 275
hyperrealism 260
hypertext 63, 71, 73, 75, 78, 79, 80, 81, 82, 325

IBM 20, 135
Watson Research Center 57
iBooks 109

ICA *see* Institute of Contemporary Art
icons 29, 98, 99
IEEE Xplore 39
Illustrator 2, 20, 24, 39, 44, 50, 58, 124, 144, 179, 205, 208, 217, 247, 300, 302, 305, 306, 325
  CS3 304
ILM *see* Industrial Light and Magic
image editing 93
image map 178
image processing 132, 134
ImageMagic 211
Imaginary Forces 246, 257, 258
Immortal 259
iMovie 124
import 296–307
InDesign 24, 75, 205, 302, 331
Industrial Light and Magic 145–6, 147
industrial media 92
Inferno 246
information 133
information processing 132
information society 100
information visualization 65, 115
infovis 115, 118
Ingalls, Dan 100
Inkscape 50
Innis, Robert 10
Inscape 217
Instagram 184
Institute of Contemporary Art (ICA) 63
Intel 7
interactive applications 167
Interactive Flash 80
Interactive Generative Stage 187
interactive kiosks 166
interactive narrative film 161
interactivity 75, 166–7
interface principles 99
Internet 62, 147, 161, 332
  studies 205
Internet Galaxy, The 183
Interpretation of Dreams 169
Invisible Shape of Things Past, The 187, 189, 191, 196, 197, 237
iOS 7, 24, 26, 29, 182, 228, 230
  apps 108
IP addresses 8
iPad 26, 101, 109
iPhone 108, 135, 165, 234
  Human Interface guidelines 101
  OS 148
iPhoto 27, 124, 184, 209, 224
IT Magazine 299
Itsu 257
iTunes 26, 35, 109, 193

Java 218, 238
Javascript 16, 31, 105, 148, 211, 218, 238
Jewish Museum 63
journalism 205
Joyce, James 82
JPEG 34, 150, 207, 216, 220, 338

Kandinsky, Wassily 236
Kay Alan 5, 13, 55–106, 107, 109, 110, 123, 135, 140, 141, 150, 161, 162, 163, 176, 180–1, 191, 231, 233, 244, 275, 318, 325, 329, 333
Keyhole, Inc. 192
Kiesler, Frederick 314–15
Kindle 109
Kirschenbaum, Matthew 16, 42
Kittler, Friedrich A. 14, 20
KML 209
Knoll, John 146

Knoll, Thomas 146
Kodak 224
Kunsthaus, Graz 315

Lakoff, George 136
  theory of metaphor 136–7
languages 94–5, 148, 169, 170, 211, 218, 232, 275, 290
  programming 96, 98
Languages of Art 65
Laocoon 65
laptops 60, 108
laser printers 57
Last Starfighter, The 294
Last Year at Marienbad 81, 319
Latour, Bruno 16
layer comps 209
layers 277–82
LCD 153
Le Corbusier 40
Legend of Zorro, The 258
Lessig, Lawrence 15
Lessing, Gotthold Ephraim 65, 119, 236
  Laocoon 65
Lia 138
libraries 177, 220
  digital 38–9
Library of Congress 227
licenses 38
Licklider, J. C. R. 4, 13, 40, 56, 63, 83, 88, 161
LightWave 3D 203
Lincoln Laboratory 60
linguists 32
linkability 124
Linux 7
Lipschultz, Mindi 253
Lislegaard, Ann 258, 319, 322, 323
Lissitzky, El 271
Literary Darwinism 33
Louvre 166
Lovink, Geert 16
Lucas, George 156
Lucasfilm 145
Ludovico, Alessandro 14
Lumière brothers 40
Lunenfeld, Peter 16
Lüsebrink, Dirk 170
Lust Studio 300
Lynn, Greg 303, 311–12
Lyotard, Jean-François 337

Mac OS 29
  Preview 39
MacDraw 57
Macintosh 57, 101, 105, 108, 332
Mackenzie, Adrian 16
MacPaint 41
Macromedia 147, 283
Macromedia Director 166, 286
MacWrite 41, 57
MAD Architects 315
Mad Men 257
Magritte, René 319
Malevich, Kazimir 271
Malraux, André 226
Manet, Edouard 121
Manga 227
Mappr 184, 196
Marino, Mark 42
Marks, Harry 288
Marx, Karl
  theory of social development 169
mashup culture 60
mashups 190, 191, 196
  technology 191
mass media 35
Mateas, Michael 12
Mathematical Theory of Communication, A 34
Matlab 238
Matrix series 259

MAX 39, 333
Max/MSP 105
Maya 2, 24, 31, 39, 41, 75, 124, 179, 203, 206, 247, 288, 300, 302, 311, 331
McLaren, Norman 254, 266, 288
McLean, Alex 14
McLuhan, Marshall 10, 33, 84, 113
  Understanding Media 97
Mechanization Takes Command 5
media 4–6
  technologies 4
media authoring 148
media blogs 65
media combinations 178
media culture 323
media design 5, 235, 289–96
media editing 103, 122, 135, 148
media editors 99, 118
media gestalts 167
media hub 108
media hybridity 257, 267, 282, 307
media hybridization 45, 163
media hybrids 163, 167, 243
media independence 113
media industries 35
media interface 29
media manipulation 177
media platforms 143–4, 182
media processors 75
media restructuring 171
media species 233–9
media studies 205
media technologies 238
media theorists 43, 60
media theory 97, 149, 221
Median 132
MediaWiki 27
mediums 226
Memex 73
memory 123
Memory Wall 164
mental processes 150
metalanguage 170, 244, 268–9, 274, 275, 276, 277
metamedium 101–6, 110, 112, 113, 123, 125, 161, 162, 176–84, 204, 225–33, 275, 329, 337
Meyer, Chris 255
Meyer, Trish 255
Microsoft 6, 93, 287
Microsoft Bing Maps 143, 277
Microsoft Office 304
Microsoft Virtual Earth 171
Microsoft Web Office 25
Microsoft Windows 58
Microsoft Word *see* Word
Miller, Paul D. 16
Minecraft 138
MIT 164, 332
  Architecture Machine Group 63, 176, 181
  Lincoln Laboratory 60, 95, 97
  Media Lab 40, 63, 147
Mitchell, William J. 16
MK12 257, 268
MMS *see* multimedia messaging service
mobile media platforms 28, 182
mobile multimedia messages 65
mobile phones 182
Mod Lang 315
modernism 82, 83, 120, 122
  literary 82
MOMA 30, 40
monomedium 225–33
Montfort, Nick 10, 16, 42
Moretti, Franco 168
Mortitz, William 286
Morville, Peter 115
Moss, Eric 312
Mossberg, Walter S 109
Motion 246

Motion Graphics 46, 163, 169, 179, 254
Motion Painting 315
Motion Painting I 286
mouse 57, 98, 99
Moviola 206
Mozilla Foundation 27
MP3 players 153
multimedia 62, 161–76, 193, 305
  artworks 65
  documents 167
  websites 166
multimedia messaging service (MMS) 151, 165, 167
  messages 170
multiple mentalities, Bruner's theory of 98
multitrack recording 145, 281
Murata, Takeshi 258, 260–1, 266, 307, 314, 319, 323
  Untitled (Pink Dot) 260–1, 266
Murray, Janet 16
music technology 145
music video 46
MySpace 47, 334

Nanika 164
Nasdaq Composite Index 6
National Gallery, London 228
naturalism 122
Negroponte, Nicholas 5, 13, 56, 63, 83, 147, 176
Nelson, Ted 4–5, 13, 40, 56, 60, 71, 75, 78, 79, 81, 82, 83, 85, 91, 96, 102, 161, 162, 274, 325, 329
Netflix 35
Network Measurement Centre 184
neuroscience 150
new media 1, 95, 113
  art 1
  criticism 75
  theory 75
New York Port Authority Gateway 311–12
NeXT Workstation 58
Nipkow 155
NLS system 184
Nobel Field 164
Nobel Peace Center 164
noise 134
Nokia 164
North by Northwest 279
notebooks 108
Nuke 247
Number of Generators 137
NURBS 201, 204

Office of War 227
On the Origin of Species 176
online digital libraries 38
OpenLayers 191
OpenOffice 206
OpenStreetMap 38, 330
operating systems 275
options 221, 222
Oracle 9
Ordos Museum 315
Orphism 82, 83
Oulipo 81

Pacific Data Images 252
Page Discription Language 59
PageMaker 41, 44, 58, 332
Pages 206
paid and locked applications 37
Paint 91
paint systems 89
Paintbox 246
Painter 138
Painting with Light 251
Palette Knife 130
Panoramio 234
paper documents 71, 78

Papert, Seymour 56, 102
parameterization 223
parameters 219–25
PARC *see* Xerox PARC
Paul, Christine 162
Paul, Les 145
PCs 1–9, 55, 57, 58, 255
  applications 108
  graphics capabilities 115
PD 105
PDFs 34
Performing Arts Centre, Saadiyat Island 315
Perl 16, 105, 218
personal dynamic media 61
Peterson, Denis 260
Phaidon 40
Photobucket 124
Photoshop 2, 4, 21, 23, 24, 31, 32, 39, 41, 44, 49, 50, 58, 59, 60, 75, 84, 89, 92, 118, 124–47, 151, 179, 202, 205, 208, 210, 213, 216, 223, 246, 247, 289, 300, 302, 305, 306, 324, 325, 331, 338
  1.0 41
  2.0 41
  3.0 142
  5.5 47
  Add Noise filter 139
  Artistic submenu 138
  Brush Strokes filter 139
  Clouds filter 141
  commands 129, 140
  CS3 304
  CS4 125, 129, 142, 217
  CS5 210, 211
  CS5.5 209
  Elements 224
  filters 129, 130, 131, 134, 136–8, 139, 274
  Help 145
  Layer Groups 142
  Layers palette 125, 141–4, 145
  menus 125, 129, 134
  online Help 142
  Render Clouds filter 203, 210
  Texture submenu 138
  UI 210
  Wave filter 136 138, 210
Photosynth 196, 197
PHP 16, 105, 148, 218
Piaget, Jean 98
Picasa 47, 124, 209, 224, 334
  3.0 151
Pingala 154
Pinterest 24, 331
pixels 136, 151, 154, 202, 203, 207, 209, 222, 270, 332, 340
Plaid 257
platform studies 42
platforms 229–30
Pleix collective 257
plug-ins 92, 163, 202, 216, 224
PNG 220
Pogue, David 109
Porter, Thomas 145, 146
Porter, William 339
post-modernism 180, 267
Postscript 59
PowerPoint 2, 24, 27, 65, 83, 165, 169, 170
pre-computational media 94
Premiere 58, 206, 283, 302, 305, 306
Premiere import 47
pre-modernist art 250
print media 71
Pro Tools 144
pro-ams 205
Processing 16, 105, 133, 177, 238
program interfaces 179
programmers 93
programming languages 96

programming techniques 144–7
programming tools 94
programs 2, 103
Project Gutenberg 122
prosumers 205
Prototype theory 231
psychology 97, 98
Psyop 257, 258
Ptushko, Aleksandr 286
publishers 40
PubMed 39
Puckette, Miller 105
Puerta America hotel, Madrid 164
pull-down menus 269
Pure Data 177
Python 16, 105, 238

Quantel Paintbox 251
Quay brothers 286
QuickTime 27, 58, 84, 153, 164, 166
  1.0 164

R/Greenberg 254
radio 156
RAM 32
Rambo: First Blood 266
Rauschenberg, Robert 278
Raw formats 338–9
RCA 217
Reas, Casey 105
RED Entertainment System 230
Reduce Noise 132
Reebok-I-pump 258
remediation 59
remix 46, 60, 167–8, 269
remixability 267–77
remixing 122, 163
reproduction technologies 155
Research Centre for Augmenting Human Intellect 63, 72
Resnais, Alain 319
resolution 299
Revit 247
Rhapsody 35
Rheingold, Howard 13, 232
Rizzoli 40
Roberts, Larry 161
Rodriguez, Robert 259
RSS 329, 334
Russian Formalism 33

S&P 500 Index 6
Sack, Warren 12
Safari 47
SAGE 95, 97
Salen, Katie 16
San Francisco, Museum of Modern Art 234
Sanger, Larry 161
satellite imagery 134
Sauter, Joachim 170, 187, 192
  Invisible Shape 237
Schultz, Pit 14
Science Direct 39
SciVerse 39
Scopus 39
Scorsese, Martin 259
Scott de Martinville, Edouard-Léon 154
Scribd 2
scripts 211
scrolling 104
SeaMonkey 27
search 114
search engines 7, 114
searchability 114, 119, 122, 124, 151
Second Life 165
Sengers, Phoebe 12
sexual reproduction 167–8
Shannon, Claude 34
Shannon's Information Theory 228
Sharpen 132
Shoup, Richard 89

SuperPaint 89
Shulgin, Alexei 14
SIGGRAPH 146, 339
Silicon Graphics 148
Silicon Valley 32, 40
SimCity 329
Sims, The 329
Sims, Zach 17
simulated media 113
simulations 104
Sin City 259
Sketchpad 44, 47, 63, 86, 88, 93, 95, 96, 97, 107, 112, 135, 161, 325
Skype 27
Small Design 164
Smalltalk 63, 98, 99, 100, 103
Smart TV App 26
Smith, Alvy Ray 88–9
Snyder, Zack 259
social media 28, 46–7, 84, 205
Social Network 30
social networking 28
social networks 46, 50
sociologists 32
Sodium Fox 260–1, 266, 315, 318, 319
Softness 130
software artists 31
Software Studies Workshop 11
SoundEdit 58
sounds 29
Spirit, The 259
Spuybroek, Lars 312
SRI International 184
Stalker 319
Stamen Design 184
Star Trek II: The Wrath of Khan 146
Starewicz, Wladyslaw 286
Steenbeck 206
Stein, Jeff 280
Sterling, Bruce 16
storage 134
Stroke Detail 130
Stroke Size 130
Stylize 130
Supermodernism 275
SuperPaint 89, 90
Superstudio 274
Suprematism 82, 83
Surrealism 83
Surrealists 82, 319
Sutherland, Ivan 4, 40, 44, 47, 56, 60, 72, 86, 88, 91, 93, 95, 96, 97, 102, 107, 113, 135, 147, 162, 325, 329, 332
Svankmajer, Jan 286

tablets 182
Tarkovsky, Andrei 319
Taylor, Bob 13, 161
television 89–90, 156
Terminator 2 58, 259
Terravision 192
Tesler, Larry 100, 114
Thacker, Chuck 100
Theremin, Léon 155
thermodynamics 169
Thompson, Kristin 276
Thomson Reuters 257
3D 289–96
3D NURBS model 209
3D polygonal model 209
3D printing 93
3ds Max 2, 247, 302
3ds Max Studio 203
TIFF 220
Titanic 259
touch screens 29
Touring 193
transparency 277–82
Trnka, Jiří 286
Tron 294
Tumblr 24
Turing, Alan 59, 70, 94, 135

TV commercials 259
Twitter 6, 7, 24, 37, 47, 329
  apps 24
Twitter API 190
TX-2 computer 97

UCLA
  School of Engineering and Applied Science 184
UI *see* User Interface
UN Studio 312
Understanding Media 97, 113
United Nations
  Millennium Development Goals Monitor 37
Universal Turing Machine 70
USB 334
User Interface (UI) 98, 215, 219, 269, 285, 286
Utah, University of 63

V Australia 230
VBScript 211
VCR 170
vector image 209
vector shapes 47
Velvet Revolution 253, 254, 255, 277, 280, 288, 291, 323, 324, 325
Vertigo 279
vibrating surfaces 29
vibration feedback 29
video 156, 170
video games 139, 179
VideoWorks 58, 166
view control 75
Vimeo 24, 47, 124, 334
Virgin America 230
Virilio, Paul 16
virtual camera 179
virtual environments 179
virtual reality 75, 113
visual culture 5
visual design 45
visual languages 4
visualization 119, 151
  tools 118
Von Neumann, John 59, 94
Voyager Company 80
VRML 165
Vuitton, Louis 300, 303

Wachowski, Andy 259
Wachowski, Larry 259
Wales, Jimmy 161
Ward, Adrian 14
Wardrip-Fruin, Noah 10, 16, 42, 103, 104
Watchmen 259
Wattenberg, Martin 118
Wavefront 148, 311, 312, 324
Wax: Or the Discovery of Television Among the Bees 80
Weaver, Warren 34
Web 2.0 37
  sites 85
web applications 31
web browsers 50
web design 178
web email 50
web pages 8, 65, 170
websites 31
Weekend 279
Weibel, Peter 14
Weinbren, Graham 161
West, Kanye 257, 268
Whitney Museum of American Art 14
Whitney, James 288
Whitney, John 254, 288
Wi-Fi 334
Wikipedia 23, 27, 37, 71, 84, 118, 161, 162, 190, 234, 248, 255, 330
WiMax 334

Winchester Trilogy 315, 319
Wind 130
Windows 7, 29, 58, 98
Media Player 153
Wirth, Niklaus 207
Wittgenstein, Ludwig 231
Woolf, Virginia 82
Word 2, 4, 24, 27, 41, 44, 58, 73, 141, 155, 165, 206, 225
  dictionary 164
  Formatting Palette 104
word processors 93, 95
WordPress 2, 23, 24, 206, 334
World Wide Web 21, 35, 56, 78–9, 148, 161, 169, 182, 252
WorldMap 38
WYSIWYG 57

Xerox 85
Xerox PARC 40, 44, 57, 58, 65, 71, 72, 84, 89, 96, 98, 99, 100, 102, 103, 104, 105, 109, 114, 212, 225, 233, 325
  Learning Research Group 56, 61, 64, 100
Xerox Star 212–13
XML 201, 209

Yahoo 7
Yokohama International Port Terminal 314, 315
Your Painting 227, 228
YouTube 2, 23, 24, 25, 30, 31, 84, 124, 190, 331, 334

Zappa, Frank 145
Zeman, Karel 279
Zielinski, Siegfried 16
Zimmerman, Eric 16
Zola, Emile 121